COMPLEX MANIFOLDS

COMPLEX MANIFOLDS

JAMES MORROW
KUNIHIKO KODAIRA

AMS CHELSEA PUBLISHING
American Mathematical Society • Providence, Rhode Island

2000 *Mathematics Subject Classification.* Primary 32Qxx.

Library of Congress Cataloging-in-Publication Data

Morrow, James A., 1941–
    Complex manifolds / James Morrow, Kunihiko Kodaira.
      p. cm.
    Originally published: New York: Holt, Rinehart and Winston, 1971.
    Includes bibliographical references and index.
    ISBN 0-8218-4055-X (alk. paper)
    1. Complex manifolds.   I. Kodaira, Kunihiko, 1915– II. Title.

QA331.M82 2005
515′.946—dc22
                            2005053671

Preface

The study of algebraic curves and surfaces is very classical. Included among the principal investigators are Riemann, Picard, Lefschetz, Enriques, Severi, and Zariski. Beginning in the late 1940s, the study of abstract (not necessarily algebraic) complex manifolds began to interest many mathematicians. The restricted class of Kähler manifolds called Hodge manifolds turned out to be algebraic. The proof of this fact is sometimes called the Kodaira embedding theorem, and its proof relies on the use of the vanishing theorems for certain cohomology groups on Kähler manifolds with positive lines fundles proved somewhat earlier by Kodaira. This theorem is analogous to the theorem of Riemann that a compact Riemann surface is algebraic.

This book is a revision and organization of a set of notes taken from the lectures of Kodaira at Stanford University in 1965–1966. One of the main points was to give the original proof of the Kodaira embedding theorem. There is a generalization of this theorem by Grauert. Its proof is not included here.

Beginning in the mid-1950s Kodaira and Spencer began the study of deformations of complex manifolds. A great deal of this book is devoted to the study of deformations. Included are the semicontinuity theorems and the local completeness theorem of Kuranishi. There has also been a great deal accomplished on the classification of complex surfaces (complex dimension 2). That material is not included here.

The outline is roughly as follows. Chapter 1 includes some of the basic ideas such as surgery, quadric transformations, infinitesimal deformations, deformations. In Chapter 2, sheaf cohomology is defined and some of the completeness theorems are proved by power series methods. The de Rham and Dolbeault theorems are also proved. In Chapter 3 Kähler manifolds are studied and the vanishing and embedding theorems are proved. In Chapter 4 the theory of elliptic partial differential equations is used to study the semi-continuity theorems and Kuranishi's theorem.

It will help the reader if he knows some algebraic topology. Some results from elliptic partial differential equations are used for which complete references are given. The sheaf theory is self-contained.

We wish to thank the publisher for patience shown to the authors and Nancy Monroe for her excellent typing.

James A. Morrow
Kunihiko Kodaira

Seattle, Washington
January 1971

v

Contents

Complex Manifolds

[1]

Definitions and Examples of Complex Manifolds

I.   Holomorphic Functions

The facts of this section must be well known to the reader. We review them briefly.

DEFINITION 1.1.   A complex-valued function $f(z)$ defined on a connected open domain $W \subseteq \mathbb{C}^n$ is called *holomorphic*, if for each $a = (a_1, \cdots, a_n) \in W$, $f(z)$ can be represented as a convergent power series

$$\sum_{\substack{k_1 \geq 0, k_n \geq 0}}^{+\infty} c_{k_1 \cdots k_n}(z_1 - a_1)^{k_1} \cdots (z_n - a_n)^{k_n}$$

in some neighborhood of a.

REMARK.   If $p(z) = \sum c_{k_1 \cdots k_n}(z_1 - a_1)^{k_1} \cdots (z_n - a_n)^{k_n}$ converges at $z = w$, then $p(z)$ converges for any z such that $|z_k - a_k| < |w_k - a_k|$ for $1 \leq k \leq n$.

Proof.   We may assume $a = 0$. Then there is a constant $C > 0$ such that for all coefficients $c_{k_1 \cdots k_n}$,

$$|c_{k_1 \cdots k_n} w_1^{k_1} \cdots w_n^{k_n}| \leq C.$$

Hence

$$|c_{k_1 \cdots k_n} z_1^{k_1} \cdots z_n^{k_n}| \leq C \left| \frac{z_1}{w_1} \right|^{k_1} \cdots \left| \frac{z_n}{w_n} \right|^{k_n}. \tag{1}$$

If $|z_i/w_i| < 1$ for $1 \leq i \leq n$, (1) gives

$$\sum |c_{k_1 \cdots k_n} z_1^{k_1} \cdots z_1^{k_n}| \leq C \prod_{i=1}^{n} \left(\frac{1}{1 - \left| \dfrac{z_i}{w_i} \right|} \right) < +\infty. \qquad \text{Q.E.D.}$$

We have the following picture:

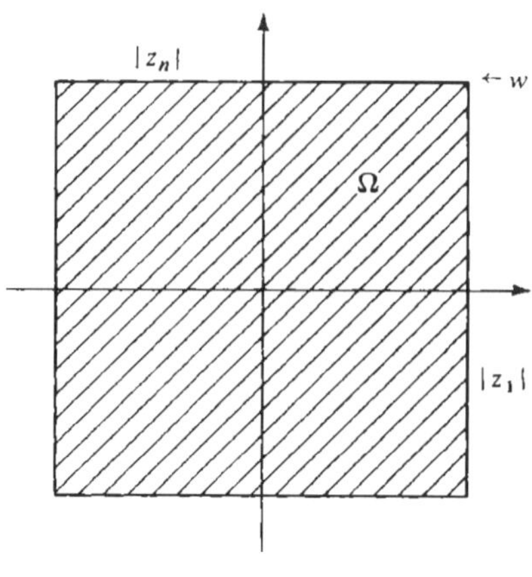

Figure 1

Ω is the region $\{z|\,|z_i| < |w_i|\, i \le n\}$.

For convenience, we let

$$P(a, r) = \{z|\,|z_\nu - a_\nu| < r_\nu,\, \nu = 1, \cdots, n\}.$$

Sometimes we call $P(a, r)$ a *polydisc* or *polycylinder*. A complex-valued function $f(z) = f(x_1 + iy_1, \cdots, x_n + iy_n)$, where $i = \sqrt{-1}$ can be considered as a function of $2n$ real variables. Then:

DEFINITION 1.2.   A complex-valued function of n complex variables is continuous or differentiable if it is continuous or differentiable when considered as a function of $2n$ real variables.

We have:

THEOREM 1.1.   (Osgood)   If $f(z) = f(z_1, \cdots, z_n)$ is a continuous function on a domain $W \subseteq \mathbb{C}^n$, and if f is holomorphic with respect to each z_k when the other variables z_i are fixed, then f is holomorphic in W.

Proof.   Take any $a \in W$ and choose r so that $\overline{P(a, r)} \le W$. We use the Cauchy integral theorem for the representation for $z \in P(a, r)$

$$f(z_1, \cdots, z_n) = \frac{1}{2\pi i} \int_{|w_1 - a_1| = r_1} \frac{f(w_1, z_2, \cdots, z_n)}{w_1 - z_1}\, dw_1,$$

$$f(w_1, z_2, \cdots, z_n) = \frac{1}{2\pi i} \int_{|w_2 - a_2| = r_2} \frac{f(w_1, w_2, z_3, \cdots, z_n)}{w_2 - z_2}\, dw_2,$$

and so on.

Substituting we get

$$f(z) = \left(\frac{1}{2\pi i}\right)^n \int_{|w_1 - a_1| = r_1} \cdots \int_{|w_n - a_n| = r_n} \frac{f(w_1, \cdots, w_n)}{(w_1 - z_1) \cdots (w_n - z_n)} \, dw_1 \cdots dw_n.$$

We are assuming

$$\left|\frac{z_\nu - a_\nu}{w_\nu - a_\nu}\right| < 1.$$

Hence the series

$$\frac{1}{w_\nu - z_\nu} = \frac{1}{(w_\nu - a_\nu) + (a_\nu - z_\nu)} = \left[\frac{1}{1 - (z_\nu - a_\nu/w_\nu - a_\nu)}\right] \frac{1}{w_\nu - a_\nu}$$

$$= \left(\frac{1}{w_\nu - a_\nu}\right) \sum_{k=0}^{\infty} \left(\frac{z_\nu - a_\nu}{w_\nu - a_\nu}\right)^k$$

converges absolutely in $P(a, r)$. Integrating term by term we get

$$f(z) = \sum_{n=0}^{\infty} c_{k_1} \cdots k_n (z_1 - a_1)^{k_1} \cdots (z_n - a_n)^{k_n}, \tag{2}$$

where

$$c_{k_1} \cdots k_n = \left(\frac{1}{2\pi i}\right)^n \int_{|w_1 - a_1| = r_1} \cdots \int_{|w_n - a_n| = r_n} \frac{f(w_1 \cdots, w_n) \, dw_1 \cdots dw_n}{(w_1 - a_1)^{k_1 + 1} \cdots (w_n - a_n)^{k_n + 1}}.$$

Then

$$|c_{k_1} \cdots k_n| \leq M \frac{1}{r_1^{k_1} \cdots r_n^{k_n}},$$

where $M = \sup\{|f(w)| \,|\, w \in \overline{P(a, r)}\}$. It follows that the representation (2) for $f(z)$ is valid for $z \in P(a, r)$ and hence the theorem is true.

We now introduce the Cauchy-Riemann equations. Let $f(z)$ be a differentiable function on domain $\Omega \subseteq \mathbb{C}^n$.

DEFINITION 1.3.   The operators $\partial/\partial z_\nu$, $\partial/\partial \bar{z}_\nu$, $1 \leq \nu \leq n$ are defined by

$$\frac{\partial f}{\partial z_\nu} = \frac{1}{2}\left(\frac{\partial f}{\partial x_\nu} - i\frac{\partial f}{\partial y_\nu}\right),$$

$$\frac{\partial f}{\partial \bar{z}_\nu} = \frac{1}{2}\left(\frac{\partial f}{\partial x_\nu} + i\frac{\partial f}{\partial y_\nu}\right),$$

where $z_\nu = x_\nu + iy_\nu$ as usual.

Let $f(z) = u(x, y) + iv(x, y)$. Then

$$\frac{\partial f}{\partial \bar{z}} = \frac{1}{2}\left[\frac{\partial u}{\partial x} + i\frac{\partial v}{\partial x} + i\left(\frac{\partial u}{\partial y} + i\frac{\partial v}{\partial y}\right)\right]$$

$$= \frac{1}{2}\left[\frac{\partial u}{\partial x} - \frac{\partial v}{\partial y} + i\left(\frac{\partial v}{\partial x} + \frac{\partial u}{\partial y}\right)\right].$$

So, $\partial f/\partial \bar{z} = 0$ if and only if $\partial u/\partial x = \partial v/\partial y$ and $\partial v/\partial x = -\partial u/\partial y$ (the Cauchy-Riemann equations).

REMARK.   If $\partial f/\partial \bar{z} = 0$, then $df/dx = \partial f/\partial z$, where $df/dx = \partial u/\partial x + i(\partial v/\partial x)$. The following calculation verifies this:

$$\frac{\partial f}{\partial z} = \frac{1}{2}\left[\frac{\partial u}{\partial x} + i\frac{\partial v}{\partial x} - i\left(\frac{\partial u}{\partial y} + i\frac{\partial v}{\partial y}\right)\right]$$

$$= \frac{1}{2}\left[\frac{\partial u}{\partial x} + i\frac{\partial v}{\partial x} + i\left(\frac{\partial v}{\partial x} - i\frac{\partial u}{\partial x}\right)\right].$$

THEOREM 1.2.   Let $f(z)$ be a (continuously) differentiable function on the open set $\Omega \subseteq \mathbb{C}^n$. Then $f(z)$ is holomorphic if and only if $\partial f/\partial \bar{z}_v = 0$, $i \leq v \leq n$.

Proof.   This follows easily from Osgood's theorem and the classical fact for functions of one complex variable. We need another simple calculation. From now on differentiable will mean having continuous derivatives of all orders (C^∞).

PROPOSITION 1.1.   Suppose $f(w) = f(w_1, \cdots, w_m)$ and $g_\lambda(z)$ $1 \leq \lambda \leq m$ are differentiable and such that the domain of f contains the range of $(g_1, \cdots, g_\lambda) = g$. Then $f[g_1(z), \cdots, g_m(z)]$ is differentiable and if $w_\lambda(z) = g_\lambda(z)$,

$$\frac{\partial f}{\partial \bar{z}_v} = \sum_{\lambda=1}^{m}\left(\frac{\partial f}{\partial w_\lambda}\frac{\partial w_\lambda}{\partial \bar{z}_v} + \frac{\partial f}{\partial \bar{w}_\lambda}\frac{\partial \bar{w}_\lambda}{\partial \bar{z}_v}\right), \tag{3}$$

$$\frac{\partial f}{\partial z_v} = \sum_{\lambda=1}^{m}\left(\frac{\partial f}{\partial w_\lambda}\frac{\partial w_\lambda}{\partial z_v} + \frac{\partial f}{\partial \bar{w}_\lambda}\frac{\partial \bar{w}_\lambda}{\partial z_v}\right). \tag{4}$$

Proof.   All statements follow trivially from the chain rule of calculus. For punishment we calculate (3). Let $w_\lambda = u_\lambda + iv_\lambda = g_v(z)$. Then

$$\frac{\partial f[g(z)]}{\partial \bar{z}_v} = \sum_{\lambda=1}^{m}\frac{\partial f}{\partial u_\lambda}\frac{\partial u_\lambda}{\partial \bar{z}_v} + \frac{\partial f}{\partial v_\lambda}\frac{\partial v_\lambda}{\partial \bar{z}_v}.$$

Making the substitutions,

$$u_\lambda = \frac{1}{2}(g_\lambda + \bar{g}_\lambda), \qquad v_\lambda = \frac{1}{2i}(g_\lambda - \bar{g}_\lambda),$$

we get

$$\begin{aligned}
\frac{\partial f[g(z)]}{\partial \bar{z}_v} &= \sum_{\lambda=1}^{m} \left\{ \frac{\partial f}{\partial u_\lambda} \frac{1}{2}\left(\frac{\partial g_\lambda}{\partial \bar{z}_v} + \frac{\partial \bar{g}_\lambda}{\partial \bar{z}_v}\right) \right. \\
&\qquad\qquad \left. + \frac{\partial f}{\partial v_\lambda}\left(\frac{1}{2i}\right)\left(\frac{\partial g_\lambda}{\partial \bar{z}_v} - \frac{\partial \bar{g}_\lambda}{\partial \bar{z}_v}\right)\right\} \\
&= \sum_{\lambda=1}^{m} \left\{ \frac{1}{2}\left(\frac{\partial f}{\partial u_\lambda} - i\frac{\partial f}{\partial v_\lambda}\right)\frac{\partial g_\lambda}{\partial \bar{z}_v} \right. \\
&\qquad\qquad \left. + \frac{1}{2}\left(\frac{\partial f}{\partial u_\lambda} + i\frac{\partial f}{\partial v_\lambda}\right)\frac{\partial \bar{g}_\lambda}{\partial \bar{z}_v}\right\},
\end{aligned}$$

which gives (2).

COROLLARY 1.   If $f(w)$ is holomorphic in w and if $w = g(z) = [g_1(z), \cdots,$ $g_m(z)]$ where each $g_\lambda(z)$ is holomorphic in z, then $f[g(z)]$ is holomorphic in z.

COROLLARY 2.   The set $\mathcal{O}_\Omega$ of all functions holomorphic on Ω forms a ring.

In order to study complex manifolds we must consider holomorphic *maps*. Let U be a domain in $\mathbb{C}^n$ and let f be a map from U into $\mathbb{C}^m$,

$$f(z_1, \cdots, z_n) = [f_1(z), \cdots, f_m(z)].$$

DEFINITION 1.4. f is *holomorphic* if each f_λ is holomorphic. The matrix

$$\begin{pmatrix} \dfrac{\partial f_1}{\partial z_1} & \cdots & \dfrac{\partial f_m}{\partial z_1} \\ \vdots & & \\ \dfrac{\partial f_1}{\partial z_n} & \cdots & \dfrac{\partial f_m}{\partial z_n} \end{pmatrix} = \left(\frac{\partial f_\lambda}{\partial z_v}\right)_{\substack{\lambda=1,\cdots,m \\ v=1,\cdots,n}}$$

is called the *Jacobian matrix*. If $m = n$, the determinant, $\det(\partial f_\lambda / \partial z_v)$ is called the *Jacobian*. Writing out the real and imaginary parts $w_\lambda = u_\lambda + iv_\lambda = f_\lambda$, $z_v = x_v + iy_v$, we have $2n$ functions u_λ, v_λ of $2n$ real variables x_v, y_v. We write briefly

$$\det\left[\frac{\partial(u_1, v_1, \cdots, u_n, v_n)}{\partial(x_1, y_1, \cdots, x_n, y_n)}\right] = \frac{\partial(u, v)}{\partial(x, y)}.$$

REMARK.   If f is holomorphic, $\partial(u, v)/\partial(x, y) = |\det(\partial f_\lambda/\partial z_\nu)|^2 \geq 0$.

Proof.   We write it out for $n = 2$ and leave the general case to the reader. We use the Cauchy-Riemann equations and set $a_{\nu\lambda} = \partial u_\lambda/\partial x_\nu = \partial v_\lambda/\partial y_\nu$, $b_{\nu\lambda} = \partial v_\lambda/\partial x_\nu = -\partial u_\lambda/\partial y_\nu$. Then

$$\begin{vmatrix} \dfrac{\partial u_1}{\partial x_1} & \dfrac{\partial v_1}{\partial x_1} & \dfrac{\partial u_2}{\partial x_1} & \dfrac{\partial v_2}{\partial x_1} \\[2mm] \dfrac{\partial u_1}{\partial y_1} & \dfrac{\partial v_1}{\partial y_1} & \dfrac{\partial u_2}{\partial y_1} & \dfrac{\partial v_2}{\partial y_1} \\[2mm] \vdots & & & \end{vmatrix} = \begin{vmatrix} a_{11} & b_{11} & a_{12} & b_{12} \\ -b_{11} & a_{11} & -b_{12} & a_{12} \\ a_{21} & b_{21} & a_{22} & b_{22} \\ -b_{21} & a_{21} & -b_{22} & a_{22} \end{vmatrix}.$$

We perform the following sequence of operations: Multiply column 2 by i and add it to column 1; do the same with columns 4 and 3. Then multiply row 1 by i and subtract it from row 2; do the same with rows 3 and 4. Making use of the fact that $g_{\nu\lambda} = \partial f_\lambda/\partial z_\nu = a_{\nu\lambda} + ib_{\nu\lambda}$, we get

$$\frac{\partial(u, v)}{\partial(x, y)} = \begin{vmatrix} g_{11} & g_{12} & * & * \\ g_{21} & g_{22} & * & * \\ 0 & 0 & \bar{g}_{11} & \bar{g}_{12} \\ 0 & 0 & \bar{g}_{21} & \bar{g}_{22} \end{vmatrix} = |\det(g_{\nu\lambda})|^2$$

by interchanging columns 2 and 3 and rows 2 and 3.   Q.E.D.

THEOREM 1.3.   (Inverse Mapping Theorem)   Let $f: U \to \mathbb{C}^n$ be a holomorphic map. If $\det(\partial f_\mu/\partial z_\nu)|_{z=a} \neq 0$, then for a sufficiently small neighborhood N of a, f is a bijective map $N \to f(N)$; $f(N)$ is open and $f^{-1}|_{f(N)}$ is holomorphic on $f(N)$.

Proof.   The remark gives $\partial(u, v)/\partial(x, y) \neq 0$ at a. We then use the inverse mapping theorem for differentiable (real variable) functions to conclude that $f(N)$ is open, f is bijective, and f^{-1} is differentiable on $f(N)$. Set $\varphi(w) = f^{-1}(w)$; then $z_\mu = \varphi_\mu[f(z)]$. Computing,

$$0 = \frac{\partial z_\mu}{\partial \bar{z}_\nu} = \sum_{\lambda=1}^{n} \frac{\partial \varphi_\mu}{\partial w_\lambda} \frac{\partial f_\lambda}{\partial \bar{z}_\nu} + \frac{\partial \varphi_\mu}{\partial \bar{w}_\lambda} \frac{\partial \bar{f}_\lambda}{\partial \bar{z}_\nu}$$

$$= \sum_{\lambda=1}^{n} \frac{\partial \varphi_\mu}{\partial \bar{w}_\lambda} \frac{\partial \bar{f}_\lambda}{\partial \bar{z}_\nu}.$$

But $\det(\partial \bar{f}_\lambda/\partial \bar{z}_\nu) = \overline{\det(\partial f_\lambda/\partial z_\nu)} \neq 0$. So by linear algebra, $\partial \varphi_\mu/\partial \bar{w}_\lambda = 0$ and $\varphi = f^{-1}$ is holomorphic.   Q.E.D.

COROLLARY. (Implicit Mapping Theorem)   Let f_λ, $\lambda = 1, \cdots, m$ be holomorphic on $U \subseteq \mathbb{C}^n$. Let rank $(\partial f_\lambda/\partial z_\nu) = r$ at each point z of U and suppose in fact that $\det(\partial f_\lambda/\partial z_\nu)_{\substack{\lambda \leq r \\ \nu \leq r}} \neq 0$. If $f_\lambda(a) = 0$ for $\lambda \leq m$ for some $a \in U$, then in a small neighborhood of a, the simultaneous equations,

$$f_\lambda(z_1, \cdots, z_r, z_{r+1}, \cdots, z_n) = 0,$$

have unique holomorphic solutions

$$z_\lambda = \varphi_\lambda(z_{r+1}, \cdots, z_n), \qquad \lambda \leq r.$$

For more details in this section one may consult Dieudonné (1960).

2. Complex Manifolds and Pseudogroup Structures

We assume given a paracompact Hausdorff space X which will also generally be assumed connected. We want to define what we mean by a complex structure on X (or structure of a complex manifold) which will be an obvious generalization of the concept of a Riemann surface. First we want to assume X is locally homeomorphic to a piece of $\mathbb{C}^n$.

DEFINITION 2.1.   By a *local complex coordinate* on X we mean a topological homeomorphism $z: p \to z(p) \in \mathbb{C}^n$ of a domain $U \subseteq X$. $z(p) = [z^1(p), \cdots, z^n(p)]$ are the local coordinates of X.

DEFINITION 2.2.   By a system of local complex analytic coordinates on X we mean a collection $\{z_j\}_{j \in I}$ (for some index set I) of local complex coordinates $z_j: U_j \to \mathbb{C}^n$ such that:

(1) $X = \bigcup_{j \in I} U_j$.

(2)   The maps $f_{jk}: z_k(p) \to z_j(p)$ are biholomorphic [that is, $z_j \circ z_k^{-1} = f_{jk}$ and $f_{jk}^{-1} = z_k \circ z_j^{-1}$ are holomorphic maps from $z_k(U_j \cap U_k)$ onto $z_j(U_j \cap U_k)$] for each pair of indices (j, k) with $U_j \cap U_k \neq \phi$.

DEFINITION 2.3.   Two systems $\{z_j\}_{j \in I}$, $\{w_\lambda\}_{\lambda \in \Lambda}$ are equivalent if the maps $z_j(p) \to w_\lambda(p)$ are biholomorphic when and where defined.

DEFINITION 2.4.   By a *complex structure on* X we mean an equivalence class of systems of local complex (analytic) coordinates on X. By a *complex manifold M* we mean a paracompact Hausdorff space X together with a complex structure defined on X.

EXAMPLE. Complex projective space $\mathbb{P}^n$. This is constructed from $\mathbb{C}^{n+1} - \{0\}$ by identifying $(p \sim q)p = (p^0, p^1, \cdots, p^n)$ and $q = (q^0, \cdots, q^n)$ if and only if $p^\lambda = cq^\lambda$ for some nonzero $c \in \mathbb{C}$, for $0 \le \lambda \le n$. Then $\mathbb{P}^n = \mathbb{C}^{n+1} - \{0\}/\sim$ is a compact Hausdorff space and one can construct a system of complex coordinates as follows: We let $U_j = \{p \in \mathbb{P}^n | p^j \ne 0\}$. Then $\{U_j\}_{j \le n}$ is an open covering of $\mathbb{P}^n$. On U_j the map $z_j = (z_j^0, \cdots, z_j^{j-1}, z_j^{j+1}, \cdots, z_j^n)$, where $z_j^\lambda = p^\lambda/p^j$ gives a local coordinate on U_j; in fact, $z_j(U_j) = \mathbb{C}^n$. Then $f_{jk} \colon z_k \to z_j$ is given by $z_j^\lambda = z_k^\lambda/z_k^j$ for $\lambda \ne k$, $z_j^k = 1/z_k^j$. (One simply multiplies by p^k/p^j.) Thus we see that $\{U_j, z_j\}$ is a complex analytic system defining a complex structure on $\mathbb{P}^n$.

Generalizing this procedure we introduce the idea of a pseudogroup structure. All spaces will be Hausdorff in what follows.

DEFINITION 2.5.   A *local homeomorphism f* between two spaces X and Y is a homeomorphism of an open set U in X to an open set $f(U)$ in Y. One has a similar definition of local diffeomorphism. A local homeomorphism (diffeomorphism) of X is such a map with $X = Y$.

Let ϑ be a domain of $\mathbb{R}^n$ or $\mathbb{C}^n$. Let f and g be local diffeomorphisms of ϑ. If open $W \subseteq \vartheta$, $f|W$ denotes f restricted to W which is the restriction of f to domain $(f) \cap W$. If W is some open set such that g is defined on W and $W \cap f(U) \ne \phi$, then $g \circ f$ is defined on $f^{-1}[W \cap f(U)]$.

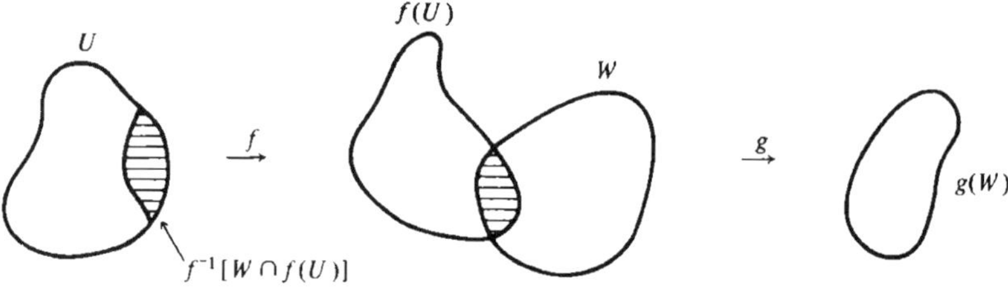

Figure 2

DEFINITION 2.6.   A *pseudogroup* of transformations in ϑ is a set Γ of local diffeomorphisms of ϑ such that

    (1) $f \in \Gamma \Rightarrow f^{-1} \in \Gamma$.
    (2) $f \in \Gamma, g \in \Gamma \Rightarrow g \circ f \in \Gamma$ where defined.
    (3) $f \in \Gamma \Rightarrow f|W \in \Gamma$ for any open $W \subseteq \vartheta$.
    (4) The identity map id $\in \Gamma$.
    (5) (completeness) Let f be any local diffeomorphism of ϑ. If $\vartheta = \cup U_j$ and $f|U_j \in \Gamma$ for each j, then $f \in \Gamma$.

DEFINITION 2.7. Let Γ (a pseudogroup on ϑ) and X (a paracompact Hausdorff space) be given. By a system of local Γ-coordinates we mean a set $\{z_j\}_{j \in I}$ of local topological homeomorphisms z_j of X into ϑ such that $z_j \circ z_k^{-1} \in \Gamma$ whenever it is defined. $\{w_\lambda\}$ and $\{z_j\}$ are equivalent (Γ-equivalent) if $w_\lambda \circ z_j^{-1} \in \Gamma$ when defined. A Γ-structure on X is an equivalence class of systems of local Γ-coordinates on X. A Γ-manifold is a paracompact Hausdorff space X together with a Γ-structure on X.

EXAMPLES

1. $\vartheta = \mathbb{C}^n$, $\Gamma_c = $ (all local biholomorphic maps of $\mathbb{C}^n$). Then a Γ_c-structure is a complex structure, and a Γ_c-manifold is a complex manifold.

2. $\vartheta = \mathbb{R}^n$, $\Gamma_d = $ (all local diffeomorphisms of $\mathbb{R}^n$). Then a Γ_d-structure is a differentiable structure and a Γ_d-manifold is a differentiable manifold.

3. Let Γ be the set of a local diffeomorphism f of $\mathbb{R}^{2n}$ satisfying the following condition. The matrix $(\varepsilon_{\lambda v})$ will be defined to be

$$\begin{pmatrix} 0 & -1 & & & & & & \\ 1 & 0 & & & & & 0 & \\ & & 0 & -1 & & & & \\ & & 1 & 0 & & & & \\ & & & & \ddots & & & \\ & & & & & & 0 & -1 \\ 0 & & & & & & 1 & 0 \end{pmatrix},$$

where the blocks $\begin{pmatrix} 0 & -1 \\ 1 & 0 \end{pmatrix}$ occur on the diagonal and the rest of the entries are zeros. If $x = (x^1, \cdots, x^{2n}) \in \mathbb{R}^{2n}, f(x) = [f_1(x), \cdots, f_{2n}(x)]$ then the derivatives of f should satisfy

$$\sum_{\lambda, v = 1}^{2n} \mathscr{E}_{\lambda v} \frac{\partial f_\lambda}{\partial x^p} \frac{\partial f_v}{\partial x^q} = \mathscr{E}_{pq}.$$

A system satisfying Example 1 is called a *Hamiltonian dynamical system*, and such an f is a *canonical transformation*. In this case a Γ-structure is called a *canonical structure*.

4. Let $\Gamma = $ (local affine transformations of $\mathbb{R}^n$). These transformations have the form

$$f^\lambda(x) = \sum_{v=1}^{n} a_v^\lambda x^v + b^\lambda,$$

where the a_v^λ, b^λ are constants and the matrix (a_v^λ) is nonsingular. In this case a Γ-structure is called *flat affine* structure.

If pseudogroup structures Γ_1 and Γ_2 are such that $\Gamma_1 \subset \Gamma_2$, then every system of local Γ_1 coordinates is a system of local Γ_2 coordinates, and Γ_1 equivalence implies Γ_2 equivalence. Hence, every Γ_1-structure determines a

Γ_2-structure. By assumption $\Gamma \subset \Gamma_d$ for all Γ. So every Γ-structure on X determines a differentiable structure on X and every Γ-manifold is a differentiable structure on X and every Γ-manifold is a differentiable manifold. The Γ-structure M is defined on the differentiable manifold X.

The problem of determining the Γ-structures on a given differentiable manifold M for given Γ is one of the most important (and difficult) problems in geometry. It is known, for example, that if X is a compact orientable differentiable surface (real dimension 2), then the only complex structures on X are those of the classical Riemann surfaces. In case $X = S^2$ (as a differentiable manifold), then $X = \mathbb{P}^1$ complex analytically (this is a classical fact). If the underlying differentiable manifold X is diffeomorphic to $\mathbb{P}^n$, then one conjectures that $X = \mathbb{P}^n$ complex analytically [see Hirzebruch and Kodaira (1957)], and Kodaira and Spencer (1958). If S^{2n} is the sphere with its usual differentiable structure, it can be shown [Borel and Serre (1953) and Wu (1952)] that S^{2n} for $n \neq 1,3$ has no complex structure

$$[S^{2n} = \{(x_1, \cdots, x_{2n+1}) \mid \sum_{i=2}^{2n+1} x_i^2, (x_1, \cdots, x_{2n+1}) \in \mathbb{R}^{2n+1}\}].$$

For S^2 there is the usual complex structure. It has been recently proved by A. Adler (1969) that S^6 has no complex structure. As a final example, let M be a compact surface and let Γ^+ be the pseudogroup of all local affine transformations,

$$x_v \rightarrow a_{v1} x_1 + a_{v2} x_2 + b_v, \qquad v = 1, 2$$

such that

$$\begin{vmatrix} a_{11} & a_{12} \\ a_{21} & a_{22} \end{vmatrix} > 0.$$

We have:

THEOREM 2.1. [Benzecri (1959)] If a Γ^+-structure exists on M, then the genus of M is 1. If M is not a torus, then M cannot be covered by any system $\{(x_j^1, x_j^2)\}$ of local coordinates such that $|\partial x_j^\lambda / \partial x_k^v|$ is constant on $U_j \cap U_k$ for each pair of indices (j, k).

The proof will not be given here.

We continue with the definitions. Let M be a complex manifold, W an open set in M, and $\{z_j\}$ a coordinate system. Then a mapping $f \colon W \to \mathbb{C}^l$ is *holomorphic* (*differentiable*, and so on) if $f \circ z_j^{-1}$ is *holomorphic* (*differentiable*, and so on) for each j where defined. Let N be another complex manifold with coordinates $\{w_\lambda\}$ and $f \colon W \to N$. Then f is *holomorphic* (*differentiable*, and so on) if $w_\lambda \circ f \circ z_j^{-1}$ is holomorphic where defined.

DEFINITION 2.8. A subset $S \subseteq M$ of a complex manifold is a (complex) analytic subvariety if, for each $s \in S$, there are holomorphic functions $f_\lambda(p)$ defined on a neighborhood $U \ni s$, $1 \leq \lambda \leq r$, such that $S \cap U = \{p \mid f_\lambda(p) = 0,$

$1 \leq \lambda \leq r\}$. Then $f_\lambda = 0$, $1 \leq \lambda \leq n$, are the local equations defining S at s. The subvariety S is called a *submanifold* if S is defined at each $s \in S$ by local equations $f_\lambda = 0$ such that

$$\operatorname{rank} \left[\frac{\partial f_\lambda(p)}{\partial z_j^\nu(p)} \right] = r \text{ is independent of } s.$$

Suppose $\det(\partial f_\lambda / \partial z_j^\nu)_{\substack{1 \leq \lambda \leq r \\ 1 \leq \nu \leq r}} \neq 0$. Then letting

$$w_j^\lambda(p) = f_\lambda(p), \qquad \text{for } \lambda = 1, \cdots, r$$

$$w_j^\lambda(p) = z_j^\lambda(p), \qquad \text{for } \lambda = r+1, \cdots, n,$$

we have a local coordinate $w_j(=(w_j^1, \cdots, w_j^n))$ such that $S: w_j^1 = w_j^2 = \cdots = w_j^r = 0$ (is defined by). Let $\zeta_j^\lambda(p) = w_j^{r+\lambda}(p) = z_j^{r+\lambda}(p)$ for $p \in S \cap U_j$. Then S is a complex manifold with local coordinates $\{\zeta_j\}$.

We want to introduce meromorphic functions on a complex manifold. They should be those functions which are locally quotients of holomorphic functions. More precisely:

DEFINITION 2.9.   A *meromorphic* function f on M is a complex-valued function defined outside of some proper subvariety S of M ($S \neq M$) and such that given $q \in M$, there is a neighborhood U of q in M and local holomorphic functions g, h on U such that $f(p) = g(p)/h(p)$ for $p \in U - S$.

EXAMPLES
1.   Any holomorphic map $f: M \rightarrow \mathbb{P}^1 = C \cup \{\infty\}$, $[S = f^{-1}(\infty)]$.
2.   In $\mathbb{C}^2$, $f(z_1, z_2) = z_1/z_2$ or $f(z_1, z_2) = P(z_1, z_2)/Q(z_1, z_2)$, where P and Q are polynomials.

3.   Some Examples of Construction (or Description) of Compact Complex Manifolds

First we have submanifolds of known manifolds ($\mathbb{P}^n$, $\mathbb{P}^n \times \mathbb{P}^n$, and so on). Let $\mathbb{P}^n$ have homogeneous coordinates $(\zeta_0, \cdots, \zeta_n)$. Let $f_\lambda(\zeta)$, $1 \leq \lambda \leq m$ be homogeneous polynomials and define $M = \{\zeta \mid f_\lambda(\zeta) = 0, 1 \leq \lambda \leq m\}$. Such an M is called a *projective algebraic* (or simply *algebraic*) variety. If the rank of $(\partial f_\lambda / \partial \zeta_\nu)_\zeta$ is independent of $\zeta \in M$, then M is a complex manifold. These are exactly the classical algebraic (projective algebraic) manifolds. In some cases the equations $f_\lambda = 0$ give some easily read information about M. For instance, if f is homogeneous of degree d, then $M_d = \{\zeta \mid f(\zeta) = 0\}$ is called a hypersurface in $\mathbb{P}^n$ of order d. If at least one of $(\partial f / \partial \zeta_\lambda)(\zeta) \neq 0$, $1 \leq \lambda \leq n$, for each $\zeta \in M_d$, then M_d is nonsingular.

EXAMPLES

1. $M_d \subseteq \mathbb{P}^2$ a nonsingular plane curve of order d is a Riemann surface of genus $g = \frac{1}{2}d(d-3) + 1$.

2. A nonsingular $M_d \subseteq \mathbb{P}^3$. M_d is simply connected and the Euler number $\chi(M_d) = d(d^2 - 4d + 6)$. [The formulas in Examples 1 and 2 can be obtained from Hirzebruch (1962), p. 91, Equation (5). They are well-known classical formulas. The simple connectivity is also well known and it follows from the Lefschetz theorems on hyperplane sections—see Milnor (1963), p. 41.]

3. Let $M \subseteq \mathbb{P}^3$ be defined by

$$M = \{\zeta \mid \zeta_1\zeta_2 - \zeta_0\zeta_3 = 0,\ \zeta_0\zeta_2 - \zeta_1^2 = 0,\ \zeta_2^2 - \zeta_1\zeta_3 = 0\}.$$

We claim that M is complex analytically homeomorphic to P^1. One can easily check that the map $\mu : \mathbb{P}^1 \to \mathbb{P}^3$ defined by $\mu(t) = (t_0^3, t_0^2 t_1, t_0 t_1^2, t_1^3)$ where $t = (t_0, t_1) \in \mathbb{P}^1$, is a biholomorphic map of $\mathbb{P}^1$ onto M.

We remark that in the cases of complex or differentiable structures, submanifolds give many examples; but for general Γ-structures one does not usually get sub Γ-structures.

Second we get *quotient spaces*.

DEFINITION 3.1. An analytic automorphism of M is a biholomorphic map of M onto M. The set of all analytic automorphisms of M forms a group g with respect to composition. Let $G \subseteq g$ be a subgroup.

DEFINITION 3.2. G is called a properly *discontinuous group* of analytic automorphisms of M if for any pair of compact subsets K_1, $K_2 \subseteq M$, the set $\{g \in G \mid gK_1 \cap K_2 \neq \phi\}$ is finite.

DEFINITION 3.3. G has no fixed points if for all $g \in G$, $g \neq 1$, g has no fixed points.

THEOREM 3.1. If G is properly discontinuous and has no fixed point, then the quotient space M/G is a complex manifold in an obvious natural manner.

Proof. We shall assume that M is connected (or a countable union of connected manifolds) and paracompact. Hence, M is σ-compact (a countable union of compact sets). Let $M/G = \{G_p \mid p \in M\}$, where $G_p = \{g(p) \mid p \in G\}$ are the orbits of $p \in M$. As notation set $M/G = M^*$, $G_p = p^*$. We shall show that given $q \in M$ we can choose a neighborhood U of q such that $p_1, p_2 \in U$, $p_1 \neq p_2$ gives $p_2^* \neq p_2^*$. In fact, there is $U \ni q$ such that $gU \cap U = \phi$ for all $g \in G$, $g \neq 1$. M is locally compact so let $U_1 \supset U_2 \supset U_3 \cdots$ be a base of relatively compact neighborhoods at q. Then $F_m = \{g \mid gU_m \cap U_m \neq \phi\}$ is a finite subset of G and $F_m \supseteq F_{m+s}, \supseteq \cdots$. If $\exists g_m \in F_m$, $g_m \neq 1$ for all m, then since each F_m is finite, $\cap F_m \ni g$, $g \neq 1$. Therefore, $gU_m \cap U_m \neq \phi$, for all m and $U_m \to q$, gives $g(q) = q$, contradicting the nonexistence of fixed points. Hence we cover M with open sets U_j such that $p_1, p_2 \in U_j$ implies $p_1^* \neq p_2^*$ and thus, $U_j \overset{*}{\to} U_j^* = \{p^* \mid p \in U_j\}$ is $1 - 1$. We give U_j^* the complex structure that U_j has. That is, if $z_j : p \to z_j(p)$ is a local coordinate on U_j, then $z_j^* : p^* \to z_j^*(p^*) = z_j(P)$ gives a local coordinate on M^*. The system $\{z_j^*\}$ then defines a complex structure on M^* and the topology of M^* is just the quotient topology for the map $M \to M^*$.   Q.E.D.

EXAMPLES

1. *Complex tori.*   Let $M = \mathbb{C}^n$. Take $2n$ vectors $\{\omega_1, \cdots, \omega_{2n}\}$, $\omega_k = (\omega_{k1}, \cdots, \omega_{kn}) \in \mathbb{C}^n$ so that the ω_j are linearly independent over $\mathbb{R}$. Let

$$G = \{g \mid g : z \to g(z) = z + \sum_{k=1}^{2n} m_k \omega_k, m_k \in \mathbb{Z}\}.$$

$T^n = \mathbb{C}^n/G$ is a (complex) torus of complex dimension n. Let $n = 1$ and arrange it so that $\omega_1 = 1$, $\omega_2 = \omega$, where the imaginary part of ω is positive. Then $T = \mathbb{C}^1/G$.

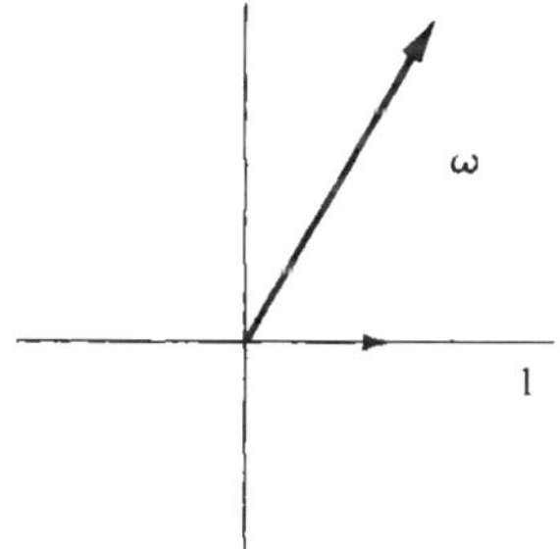

Figure 3

We have a map $\mathbb{C} \xrightarrow{\exp 2\pi i} \mathbb{C}^*$, $z \to w = e^{2\pi i z}$ where $\mathbb{C}^* = \{z \mid z \neq 0\}$. If we first take $g(z) = z + m_1\omega + m_2$ and then exponentiate, we get $e^{2\pi i(z + m_1\omega)}$. So $\exp 2\pi i \circ g = \alpha^{m_1} \cdot \exp 2\pi i$ where $\alpha = e^{2\pi i \omega}$ and $g(z) = z + m_1\omega + m_2$, and $0 < |\alpha| < 1$ since $\mathrm{Im}(\omega) > 0$. Looking a little closer we see we have the diagram

$$
\begin{array}{ccc}
\mathbb{C} & \xrightarrow{\exp 2\pi i} & \mathbb{C}^* \\
\uparrow{\scriptstyle g} & & \uparrow{\scriptstyle \alpha^{m_1}} \\
\mathbb{C} & \xrightarrow{\exp 2\pi i} & \mathbb{C}^*
\end{array}
$$

which commutes. Hence, if we let $G^* = \{g^* \mid g^* : w \to \alpha^m w, m \in \mathbb{Z}\}$, we see $T = \mathbb{C}/G = \mathbb{C}^*/G^*$.

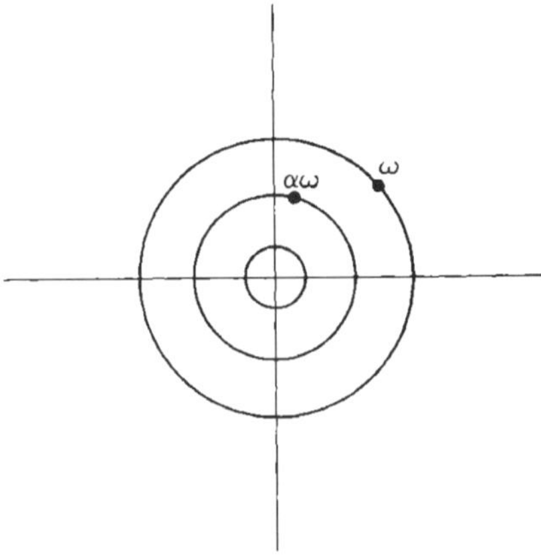

Figure 4

2. *Hopf manifolds.* Let $W = \mathbb{C}^N - \{0\}$ and $G = \{g^m \mid m \in \mathbb{Z},\ g(w_1, \cdots, w_N) = (\alpha_1 w_1, \cdots, \alpha_N w_N)$, where $0 < |\alpha_\nu| < 1\}$. Then W/G is a compact complex manifold since it is easy to see that G is properly discontinuous and has no fixed points on W. It is also easy to see that W/G is diffeomorphic to $S^1 \times S^{2N-1}$.

3. Let M be the algebraic surface (complex dimension 2) defined:

$$M = \{\zeta \mid \zeta_0^5 + \zeta_1^5 + \zeta_2^5 + \zeta_3^5 = 0\} \subseteq \mathbb{P}^3.$$

Let

$$G = \{g^m \mid m = 0, 1, 2, 3, 4 \text{ where } g(\zeta_0, \cdots, \zeta_3)$$

$$= (\rho\zeta_0, \rho^2\zeta_1, \rho^3\zeta_2, \rho^4\zeta_3) \text{ and } \rho = e^{2\pi i/5}\}.$$

Then g is a biholomorphic map $\mathbb{P}^3 \to \mathbb{P}^3$ and $g^5 = 1$. Consider the fixed points of g^m on $\mathbb{P}^3$. They satisfy $(0 = \nu \leq 3)$, $(\rho^{m(\nu+1)} - c)\,\zeta_\nu = 0$ and the fixed points are $(1, 0, 0, 0)$, $(0, 1, 0, 0)$, $(0, 0, 1, 0)$, and $(0, 0, 0, 1)$. These points are not on M so there are no fixed points on M and M/G is a complex manifold. We saw before that M is simply connected and $\chi(M) = d(d^2 - 4d + 6)$ where $d = 5$. Therefore, the Euler number of M is 55. Then the fundamental group $\pi_1(M/G) \cong G$ and $\chi(M/G) = 11$.

4. Last we have the classical examples of Riemann surfaces and their universal covering surfaces. If S is a compact Riemann surface of genus $g \geq 2$, the universal covering surface of S is the unit disk $D = \{z \in \mathbb{C}^1 \mid |z| < 1\}$. Then $S = D/G$ where each element of G is an automorphism of D and hence of the form

$$g(z) = e^{i\theta}\,\frac{z - \alpha}{\alpha z - 1}, \qquad |\alpha| < 1.$$

Finally we consider *surgeries*. Given a complex manifold M and a compact submanifold (subvariety) $S \subset M$, suppose we also have a neighborhood $W \supset S$ and manifolds $S^* \subset W^*$ with W^* a neighborhood of S^*. Suppose $f: W^* - S^* \to W - S$ is a biholomorphic map onto $W - S$. Then we can replace W by W^* and obtain a new manifold $M^* = (M - W) \cup W^*$. More precisely, $M^* = (M - S) \cup W^*$ where each point $z^* \in W^* - S^*$ is identified with $z = f(z^*)$.

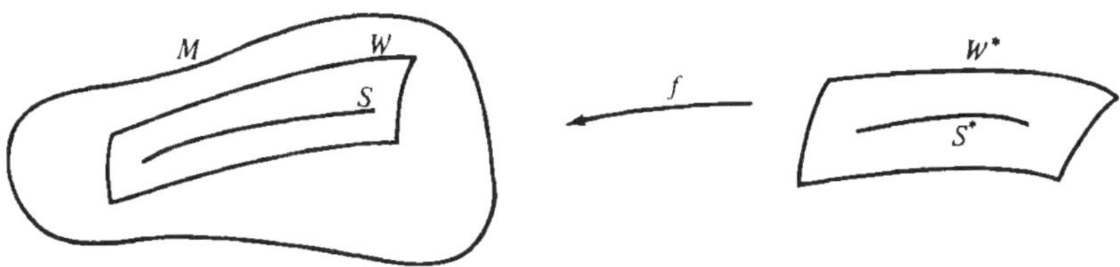

Figure 5

EXAMPLE 1.   Hirzebruch (1951) Let $M = \mathbb{P}^1 \times \mathbb{P}^1$. In homogeneous coordinates, $\mathbb{P}^1 = \{\zeta \mid \zeta = (\zeta_0, \zeta_1)\}; = \{\mathbb{C} \cup \{\infty$ in inhomogeneous coordinates, $\zeta = \zeta_1/\zeta_0 \in \mathbb{C} \cup \{\infty\}$. $M = \mathbb{P}^1 \times \mathbb{P}^1 - \{(z, \zeta) \mid z \subset \mathbb{P}^1, \zeta \in \mathbb{P}^1\}$ contains $S = \{0\} \times \mathbb{P}^1$ and $W = D \times \mathbb{P}^1$ where $D = \{z \mid |z| < \epsilon\}$ is a neighborhood of S in M. Let $W^* = D \times \mathbb{P}^{1*} = \{(z, \zeta^*) \mid z \in D, \zeta^* \in \mathbb{P}^{1*}\}$ and $S^* = \{0\} \times \mathbb{P}^*$. Fix an integer $m > 0$ and define $f: W^* - S^* \to W - S$ as follows:

$$f(z, \zeta^*) \to (z, \zeta) = [z, (\zeta^*/z^m)] \qquad \text{where } 0 < |z| < \epsilon.$$

Then f is biholomorphic on $W^* - S^*$ and let $M_m^* = (M - S) \cup W^*$ where $(z, \zeta) = (z, \zeta^*)$ if $\zeta^* = z^m \zeta$, $0 < |z| < \epsilon$.

REMARK. M and M_m^* are topologically different if m is odd.

Proof.   (for $m = 1$). $M = \mathbb{P}^1 \times \mathbb{P}^1$ is homeomorphic to $S^2 \times S^2$. We show that the homology intersection properties of M and M_1^* are distinct, hence, proving that they are topologically different. A base for $H_2(M, \mathbb{Z})$ is given by $\{S_1, S_2\}$ where $S_1 = \{0\} \times \mathbb{P}^1$, $S_2 = \mathbb{P}^1 \times \{0\}$. Hence, any 2-cycle C is homologous ($\sim$) to $aS_1 + bS_2$, $a, b \in \mathbb{Z}$. The intersection multiplicity $I(C, C) = I(aS_1 + bS_2, aS_1 + bS_2) = a^2 I(S_1, S_1) + b^2 I(S_2, S_2) + 2ab I(S_1, S_2)$. Since S_1, S_2 occur as fibres in $\mathbb{P}^1 \times \mathbb{P}^1$, $I(S_1, S_1) = I(S_2, S_2) = 0$. Hence,

$$I(C, C) = 2ab \equiv 0 \ (\text{mod } 2). \tag{1}$$

In M we have the following picture:

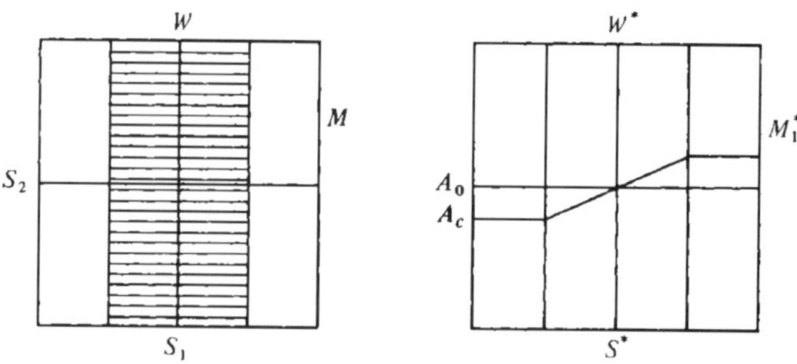

Figure 6

where A_c is the submanifold of M_1^* defined by $\zeta = c$ and $\zeta^* = zc$ with the coordinates explained before. Then A_c is a 2-cycle and $A_0 \sim A_c$. Hence $I(A_0, A_0) = I(A_0, A_c) = 1$. Since for any 2-cycle Z on M, $I(Z, Z) \equiv 0 \pmod 2$ we see $M \neq M_1^*$.

REMARKS

1. $M_m^* \neq M_n^* (m \neq n)$ as complex manifolds.
2. $M_{2m}^* = M$ topologically.
3. $M_{2m+1}^* = M_1^*$ topologically.

These facts are proved in Hirzebruch (1951).

EXAMPLE 2. (Logarithmic Transformation) Let $M = T \times \mathbb{P}^1, T = \mathbb{C}/G$, $G = \{m\omega + n \mid m, n \in \mathbb{Z}, \operatorname{Im} \omega > 0\}$ where T is a torus of complex dimension 1. For any $\zeta \in \mathbb{C}$, we denote the class in $\mathbb{C}/G = T$ by $[\zeta]$. We perform surgery on M as follows: Let $S = \{0\} \times T$, $W = D \times T$ where $\mathbb{P}^1 = \mathbb{C} \cup \{\infty\}$ and $0 \in D = \{z \in C \mid |z| < \varepsilon\}$.

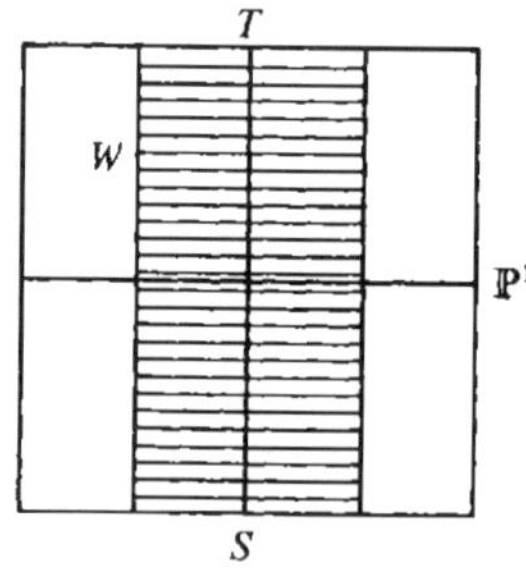

Figure 7

Then set $W^* = D \times T = \{z, [\zeta^*] \mid z \in D, [\zeta^*] \in T\}$ and $S^* = \{0\} \times T \subseteq D \times T$. Define $f: W^* - S^* \to W - S$ as follows:

$$f: (z, [\zeta^*]) \to \{z, [\zeta^* + (1/2\pi i) \log z]\},$$

where $0 < |z| < \varepsilon$.

Then f is biholomorphic and we can form $M^* = (M - S) \cup W^*$, where $(z, [\zeta]) = (z, [\zeta^*])$ if $[\zeta] = [\zeta^* + (1/2\pi i) \log z]$, $0 < |z| < \varepsilon$.

REMARK.   For the first Betti numbers b_1 we have $b_2(M) = b_2(T) = 2$, but $b_1(M^*) = 1$. In fact, M^* is topologically homeomorphic to $S^3 \times S^1$.

Proof. $H_2(M, Z) \cong Z \oplus Z$ is clear by the Kunneth theorem. To study M^*, first we notice that $M - W = (\mathbb{P}^1 - D) \times T$ is homeomorphic to $\Delta \times T$ where Δ is a closed disk, and T is homeomorphic to $S^1 \times S^1$. If $\zeta = x + y\omega$, we can identify $[\zeta]$ with (x, y), where $x + 1$ is identified with x, $y + 1$ with y, where x and y are real ($\in \mathbb{R}$). Therefore $W^* = D \times S^1 \times S^1$, $M - W = \Delta \times S^1 \times S^1$. Since we are only interested for the moment in the topological type of M^* we may as well assume that D is the unit disk and that the identification in the definition of surgery takes place on the boundary of $D = \{e^{i\theta} \mid 0 \le \theta \le 2\pi\}$. Then we identify (w, x^*, y^*) and (w, x, y) if $x = x^* + (\theta/2\pi)$, $y = y^*$. Hence, $M^* = B \times S^1$ where B is a circle bundle over S^2; and in fact, we easily see by the transition function that this is the Hopf bundle $S^3 \to S^2$. Hence $B = S^3$. This proves that $M^* = S^3 \times S^1$; $b_1(M^*) = 1$ follows.

EXAMPLE 3.   We mention also the classical *quadric transformation* (*blowing up, σ-process*). First we discuss the case where M has complex dimension 2. Let $S = p$ be any point on M, and let $S^* = \mathbb{P}^1$ be a copy of the Riemann sphere. We define $M^* = (M - p) \cup \mathbb{P}^1$ as follows. Choose a coordinate patch $W = \{(z_1, z_2) \mid |z_1| < \varepsilon, |z_2| < \varepsilon\}$ in a neighborhood of p so that $z_1(p) = z_2(p) = 0$. We define a submanifold W^* of $W \times \mathbb{P}^1$ as follows:

$$W^* = \{(z_1, z_2; \zeta_1, \zeta_2) \in W \times \mathbb{P}^1 \mid z_1\zeta_2 - z_2\zeta_1 = 0\},$$

where $(\zeta_1\zeta_2)$ are homogeneous coordinates on $\mathbb{P}^1$. W^* is a submanifold since $(\partial f/\partial z_1) = \zeta_2$, $(\partial f/\partial z_2) = -\zeta_1$ if $f = z_1\zeta_2 - z_2\zeta_1$, and hence $[(\partial f/\partial z_1), (\partial f/\partial z_2)] \ne (0, 0)$. Let $f^*: W^* \to W$ be the restriction of the projection map $W \times \mathbb{P}^1 \to W$ to W^*. Then $W^* \supseteq 0 \times \mathbb{P}^1 = S^*$, $f^*: S^* \to p = (0, 0)$, and $f^*: W^* - S^* \to W - p$ is biholomorphic. The first two statements are obvious. For the proof of the last, let $(z_1, z_2; \zeta_1, \zeta_2) \notin S^*$. Then at least one of $z_i \ne 0$ and hence (ζ_1, ζ_2) is determined by $(z_1, z_2) f^{*-1}: (z_1, z_2) \to (z_1, z_2; z_1, z_2)$. By surgery we obtain $M^* = (M - p) \cup \mathbb{P}^1$. We make the following definition:

DEFINITION 3.4.   The quadric transformation Q_p with center p is the manifold $Q_p(M) = M^*$.

REMARK. $Q_{p_m} \cdots Q_{p_1}(\mathbb{P}^2)$ can be complicated! For example,

$$Q_{p_6} \cdots Q_{p_1}(\mathbb{P}^2) = \{\zeta \mid \zeta_0^3 + \zeta_1^3 + \zeta_2^3 + \zeta_3^3 = 0\} \subseteq \mathbb{P}^3.$$

For manifolds M of dimension ≥ 2 we proceed analogously. If

$$\dim_{\mathbb{C}} M = n, \text{ let } (z_1, \cdots, z_n)$$

be coordinates centered at $p[p = (0, \cdots, 0)]$. If $W = \{(z_1, \cdots, z_n)\,|\,|z_\alpha| < \varepsilon, 1 \leq \alpha \leq n\}$, we set $W^* = \{(z, \zeta)\,|\,z_\lambda \zeta_v - z_v \zeta_\lambda = 0, i \leq \lambda, v \leq n\} \subseteq W \times \mathbb{P}^{n-1}$. Again W^* is a manifold, projection onto W defines a biholomorphic map $W^* - \mathbb{P}^{n-1} \to W - p$, by $(z, \zeta) \to z$. We form $M^* = (M - p) \cup W^* = (M - p) \cup \mathbb{P}^{n-1}$ and call $M^* = Q_p(M)$ the quadric transform of M with center p.

4.   Analytic Families; Deformations

Consider a torus $T_\omega = \mathbb{C}/G$, $G = \{m\omega + n\,|\,m, n \in \mathbb{Z} \text{ Im } \omega \to 0\}$. We have a family of tori depending on the parameter ω. Many examples of compact complex manifolds depend on parameters built into their definitions. We also have the examples of hypersurfaces of degree d in $\mathbb{P}^n$. Each such surface $M_d = \{\zeta\,|\,f(\zeta) = 0\}$ is defined by a function f of the form $f = \sum_{k_0 + \cdots + k_n = d} a_{k_0 \cdots k_n} \zeta_0^{k_0} \cdots \zeta_n^{k_n}$. In a sense to be made precise M_d depends "analytically" on the coefficients $a_{k_0 \cdots k_n}$ of f. We make the following definition:

DEFINITION 4.1. Let B be a (connected) complex manifold and let $\{M_t\,|\,t \in B\}$ be a set of *compact* complex manifolds depending on $t \in B$. We say that M_t depends *holomorphically* (or *complex analytically*) on t and that $\{M_t\,|\,t \in B\}$ forms a *complex analytic family* if there is a complex manifold $\mathcal{M}$ and a holomorphic map $\bar\omega$ onto B such that

(1) $\bar\omega^{-1}(t) = M_t$ for each $t \in B$, and
(2)   the rank of the Jacobian of $\bar\omega$ is equal to the complex dimension of B at each point of $\mathcal{M}$.

We note that (2) implies M_t is a complex submanifold of $\mathcal{M}$.

Now for some examples. As before, we denote $T_\omega = \mathbb{C}/G$,

$$G = \{n + m\omega\,|\,n, m \in \mathbb{Z}, \text{ Im } \omega > 0\}.$$

Let $B = \{\omega\,|\,\text{Im } \omega > 0\} \subset \mathbb{C}$. Let $\mathcal{G} = \{g_{mn}\,|\,g_{mn} : (\omega, z) \to (\omega, z + m\omega + n)\}$. Then $\mathcal{G}$ is a properly discontinuous group of transformations on $B \times \mathbb{C}$ without fixed point. Hence, $\mathcal{M} = B \times \mathbb{C}/\mathcal{G}$ is a complex manifold. The projection map $B \times \mathbb{C} \to B$ induces a holomorphic map $\mathcal{M} \xrightarrow{\bar\omega} B$, and $\bar\omega^{-1}(\omega) = T_\omega$. It is easy to see that the Jacobian condition is satisfied so $\{T_\omega\,|\,\omega \in B\}$ forms a complex analytic family.

But suppose we proceed as follows: Again $T_\omega = \mathbb{C}/G$ and the map $\mathbb{C} \to \mathbb{C}/G$ is written $z \to [z]$. Let $D = $ unit disk $= \{t\,|\,|t| < 1\}$. On $D \times T_\omega$ consider the group $\mathcal{G} = \{1, g\}$ where $g : (t, [z]) \to (-t, [z + \tfrac{1}{2}])$ is of order 2.

Then $\mathscr{G}$ is properly discontinuous and has no fixed points so $D \times T_\omega/\mathscr{G}$ is a complex manifold. Let $\pi\colon D \times T_\omega \to D$ be defined by $(t, [z]) \to \tau = t^2$. Then the diagram

$$
\begin{array}{ccc}
(t, [z]) & \xrightarrow{\ g\ } & (-t, [z + \tfrac{1}{2}]) \\[2pt]
\pi \downarrow & & \downarrow \pi \\[2pt]
t^2 & \xrightarrow{\ id\ } & t^2
\end{array}
$$

commutes so π defines a holomorphic map on $\mathscr{M}$. The Jacobian condition is not satisfied by π, since $(\partial\tau/\partial t) = 2t = 0$ at $t = 0$. We notice that $\pi^{-1}(\tau) = T_\omega$ if $\tau \neq 0$, but $\pi^{-1}(0) = T^*$, a torus of period $\omega/2$.

DEFINITION 4.2. Let M, N be compact complex manifolds. M is a *deformation* of N if there is a complex analytic family such that M, $N \subseteq \{M_t \mid t \in B\}$, that is, $M_{t_0} = M$, $M_{t_1} = N$.

We have the following sequence of problems to guide our work:

PROBLEM. Determine all complex structures on a given X.

PROBLEM. Determine all deformations of a given compact manifold M.

PROBLEM. (easier?) Determine all "sufficiently small" deformations of a given M.

DEFINITION 4.3. We say that all sufficiently small deformations have a certain property $\mathscr{P}$ if, for any complex analytic family $\{M_t \mid t \in B\}$ such that $M_{t_0} = M$, we can find a neighborhood N, $t_0 \in N \subset B$ such that M_t has $\mathscr{P}$ for each $t \in N$.

By standard techniques in differential topology we prove the following theorem:

THEOREM 4.1. Let M_t be a complex analytic family of complex manifolds M_t. Then M_t and M_{t_0} are diffeomorphic for any t, $t_0 \in B$.

Proof. The reader will notice that we really only use the differentiability of the map $\pi\colon \mathscr{M} \to B$, analyticity is not needed. In fact, we prove: Let $\mathscr{M}$ be a differentiable family of compact differentiable manifolds such that the differentiable map $\pi\colon \mathscr{M} \to B$ has maximal rank ($\mathscr{M}$ and B are differentiable manifolds). Then M_t is diffeomorphic to M_{t_0}.

First we construct a C^∞ vector field Θ on a neighborhood of M_{t_0} in $\mathcal{M}$ such that π induces $\pi_*(\Theta) = \partial/\partial s$, where s is a member of a coordinate system $(s, x^2, \cdots, x^m)$ in a neighborhood of the point $t_0 \in B$ chosen as follows:

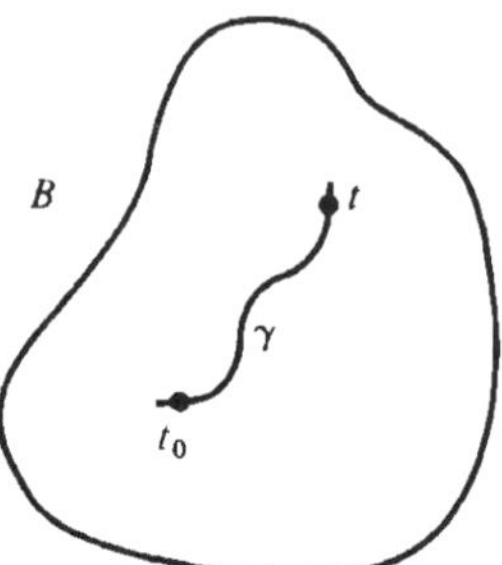

Figure 8

We connect t_0 and t by an embedded arc $\gamma: (-\varepsilon, 1 + \varepsilon) \to \{\gamma(s) \mid s \in (-\varepsilon, 1 + \varepsilon)\}$. A compactness argument shows that we can assume that t and t_0 lie in the same coordinate patch and since γ is an embedding we can find a chart with coordinate $(s, t_2, \cdots, t_m)$ around t_0 ($t_0 = (0, \cdots, 0)$, $t = (s, 0, \cdots, 0)$). Because of the rank condition, $\pi^{-1}(\gamma) = \pi^{-1}\{(s, 0, \cdots, 0) \mid -\varepsilon < s < 1 + \varepsilon\}$, is a submanifold of $\mathcal{M}$, and we can assume that $(s, x_j^2, \cdots, x_j^n)$ are coordinates of $\mathcal{M}$ for a given point of $\pi^{-1}(\gamma)$ in some neighborhood $\mathcal{U}_j$ of the point. Then the vector field $(\partial/\partial s)_j$ on $\mathcal{U}_j$ satisfies $\pi_*(\partial/\partial s)_j = \partial/\partial s$. Then if $\{\rho_j\}$ is a partition of unity subordinate to $\{\mathcal{U}_j\}$ ($\cup \mathcal{U}_j$ is a neighborhood of M_{t_0}), the vector field $\Theta = \sum_j \rho_j (\partial/\partial s)_j$ satisfies our requirements.

For the second part of the proof we seek a solution of the differential equation

$$\frac{d}{ds} x_j^\alpha(\tau) = \Theta_j^\alpha[x(\tau)], \qquad 1 \le \alpha \le n \tag{1}$$

where Θ_j^α is the α-component of Θ in the coordinate patch $\mathcal{U}_j$, with initial conditions $x_j^\alpha(0) = y^\alpha$, where $(0, y^2, \cdots, y^n)$ is some point close to $(0, \cdots, 0)$. If s is small enough and $|y|$ is small enough, Equation (1) has a unique solution $x_j(\tau, y)$ on some small interval. By compactness, we can assume that $M_{t_0} \subset \cup_j \mathcal{U}_j$, a finite union of such patches, and that in each $\mathcal{U}_j$, (1) is satisfied for $|\tau| < \mu$ where μ is independent of j. If $x_j^\alpha(\tau, y)$ is such a solution, let $x_j = f_{jk}(x_k)$ and define $\xi_k^\beta[\tau, f_{kj}(0, y)]$ uniquely on $\mathcal{U}_j \cap \mathcal{U}_k$ by

$$x_j(\tau, y) = f_{jk}(\xi_k[\tau, f_{kj}(0, y)]). \tag{2}$$

Then

$$\frac{dx_j^\alpha(\tau, y)}{d\tau} = \sum_\beta \frac{\partial x_j^\alpha}{\partial x_k^\beta} \frac{\partial \xi_k^\beta[\tau, f_{kj}(0, y)]}{\partial \tau},$$

and by the uniqueness of the solution to (1)

$$x_j(\tau, y) = f_{jk}(x_k[\tau, f_{kj}(0, y)]). \tag{3}$$

Equation (3) implies that $x(\tau, y)$, $|\tau| < \mu$, $y \in M_{t_0}$ is a well-defined differentiable map defined on M_{t_0} for each τ, $|\tau| < \mu$, and $x(0, y) = y$. Let $\varphi_\tau(y) = x(\tau, y)$; then $\varphi_0 = id$ (on M_{t_0}). It is also easy to check that $\pi[\varphi_\tau(y)] = \gamma(\tau)$ since $\pi_*(\Theta) = d/ds$. Hence, φ_τ maps M_{t_0} into $M_{\gamma(\tau)}$ (for small τ). We can repeat this argument for $M_{\gamma(\tau)}$ and define $\psi_v \colon M_{\gamma(\tau)} \to M_{\gamma(\tau + v)}$ and by uniqueness get $\psi_{-\tau} \circ \varphi_\tau = id$, $\varphi_\tau \circ \psi_{-\tau} = id$. Since everything is differentiable, the theorem is proved.   Q.E.D.

REMARK.   This argument is very old. For a treatment from the point of view of Morse theory, see Milnor (1963). Sometimes this theorem is attributed to Ehresmann (1947).

We consider some more examples of complex analytic families. The dependence of the complex structure of M_t on $t \in B$ can be complicated as we shall see.

EXAMPLE 1.   Consider again the family of tori $\{T_\omega \mid \omega \in H\}$ where $H = \{\omega \mid \mathrm{Im}\, \omega > 0\}$ and $T_\omega = \mathbb{C}/G$, $G = \{m\omega + n \mid m, n \in \mathbb{Z}\}$. From the classical theory of Riemann surfaces we see that T_ω and $T_{\omega'}$ are conformally equivalent if $\omega' = (a\omega + b/c\omega + d)$ where $a, b, c, d \in \mathbb{Z}$, and $ad - bc = 1$. Let $\mathscr{G}$ be the group of transformations acting on H which have the form

$$\omega \to \frac{a\omega + b}{c\omega + d}, \qquad a, b, c, d \in \mathbb{Z}, \qquad ad - bc = 1.$$

Then it is easily seen that $\mathscr{G}$ is properly discontinuous on H. A fundamental region $\mathscr{F}$ for $\mathscr{G}(\cup g\mathscr{F} = H$, $g\mathscr{F} \cap \mathscr{F} = \phi$ if $g \neq id)$ is given by the shaded region in the figure below, hence $T_\omega \neq T_{\omega'}$, if $\omega \neq \omega'$ and $\omega, \omega' \in \mathscr{F}$.

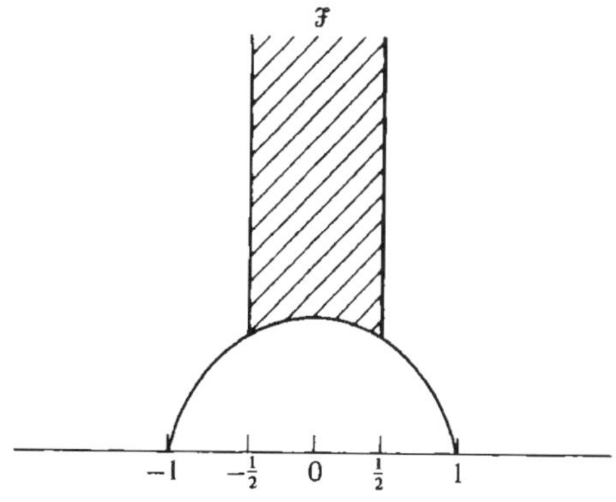

Figure 9

The elliptic modular function J defines a conformal map $J: H/\mathscr{G} \to \mathbb{C}$. So $T_\omega = T_{\omega'}$ if $J(\omega) = J(\omega')$.

EXAMPLE 2. (n-dimensional tori) We give an outline of some of the facts. A torus $T^n = \mathbb{C}^n/G$ where $G = \{\sum_{j=1}^{2n} m_j \omega_j \mid m_j \in Z\}$ where $\omega_1, \cdots, \omega_{2n}$ are complex n-vectors linearly independent over $\mathbb{R}$.

(a) We can replace $\{\omega_j\}$ by any other linearly independent basis of G. That is,

$$\omega'_{j\lambda} = \sum_{k=1}^{2n} a_{jk} \omega_{k\lambda}, \tag{4}$$

where $a_{jk} \in Z$, $\det(a_{jk}) = 1$ are also permissible generators of the lattice (group) G.

(b) We may also introduce new coordinates in $\mathbb{C}^n$ so $Z_\lambda \to \hat{Z}_\lambda$, where

$$\hat{Z}_\lambda = \sum_{v=1}^{n} Z_v \gamma_{v\lambda}, \gamma_{v\lambda} \in \mathbb{C}, \qquad \det(\gamma_{v\lambda}) \neq 0.$$

Then,

$$\hat{\omega}_{j\lambda} = \sum \omega_{jv} \gamma_{v\lambda}. \tag{5}$$

The resulting change from Equations (4) and (5) becomes

$$\hat{\omega}'_{j\lambda} = \sum a_{jk} \omega_{kv} \gamma_{v\lambda}. \tag{6}$$

We may assume that $\omega_{n+1}, \cdots, \omega_{2n}$ are $\mathbb{C}$-linearly independent. Hence by some change of coordinates $(\gamma_{v\lambda})$, we can obtain

$$\begin{pmatrix} \omega_{11} & \cdots & \omega_{1n} \\ \omega_{2n1} & \cdots & \omega_{2nn} \end{pmatrix} (\gamma_{v\lambda}) = \begin{pmatrix} \hat{\omega}_{11} & \cdots & \hat{\omega}_{1n} \\ \hat{\omega}_{n1} & \cdots & \hat{\omega}_{nn} \\ & I & \end{pmatrix}, \tag{7}$$

where I is the $n \times n$ identity matrix.

So we may assume $(\omega_{ij}) = \begin{pmatrix} \Omega \\ I \end{pmatrix}$, where $\Omega = (\omega_{ij})$ $1 \leq i, j \leq n$ and $I = \begin{pmatrix} 1 & 0 \\ 0 & 1 \end{pmatrix}$.

(c) We can also break (a_{jk}) into pieces:

$$(a_{jk}) = \begin{pmatrix} A & B \\ C & D \end{pmatrix}.$$

Then (4) takes the form

$$\omega'_{j\lambda} = (a_{jk}) \begin{pmatrix} \Omega \\ I \end{pmatrix} = \begin{pmatrix} \Omega'_1 \\ \Omega'_2 \end{pmatrix}, \qquad \Omega'_1 = A\Omega + B, \Omega'_2 = C\Omega + D.$$

If one assumes that Ω'_2 is invertible, then $\begin{pmatrix} \Omega'_1 \\ \Omega'_2 \end{pmatrix}(\Omega'_2)^{-1} = \begin{pmatrix} \Omega' \\ I \end{pmatrix}$ where

$$\Omega' = (A\Omega + B)(C\Omega + D)^{-1}, \tag{8}$$

$$\det\begin{pmatrix} A & B \\ C & D \end{pmatrix} = 1.$$

The following treatment will be a bit sketchy; for more details consult Kodaira-Spencer II (1958). The fact that $\omega_1, \cdots, \omega_n$, $(1, 0, \cdots, 0)$, $(0, 1, 0, \quad 0,), \cdots (0, \cdots, 0, 1)$ are real linearly independent implies

$$\det\begin{pmatrix} \Omega & \overline{\Omega} \\ I & I \end{pmatrix} \neq 0,$$

which is the same as $(2i)^n \det [\text{Im}(\omega_{j\lambda})] \neq 0$. Consider the space $H = \{\Omega$ $\det(\text{Im } \Omega) > 0\}$ [some sort of a generalization of $\text{Im } \omega > 0$ in Example (1)]. Let $\mathcal{G} = $ the set of all transformations

$$\Omega \to (A\Omega + B)(C\Omega + D)^{-1} = \Omega',$$

where $\begin{pmatrix} A & B \\ C & D \end{pmatrix} \in SL(n, \mathbb{Z})$, the invertible integral matrices of determinant $+1$. This group does not really act on H since it is possible for $C\Omega + D$ to be singular; one should consult Kodaira-Spencer for more details. H should be extended to something more general on which $SL(n, \mathbb{Z})$ acts. In any case,

$$T^n_\Omega = T^n_{\Omega'}, \qquad \text{if } \Omega' = g\Omega, g \in \mathcal{G}.$$

We would like to form $H/\mathcal{G}$. But it turns out that $\mathcal{G}$ is not discontinuous. In fact, for any open set $U \subset H$, there is a point $\Omega \in U$ such that $\{g\Omega \mid g \in \mathcal{G}\} \cap U$ is infinite. Hence, the topologial space $H/\mathcal{G}$ with the quotient topology is not Hausdorff and hence certainly not even a topological manifold by the usual definition.

We next give some examples of families $\{M_t \mid t \in B\}$ such that $M_t = M$ for $t \neq t_0$ and $M_{t_0} \neq M$.

EXAMPLE 3.   A *Hopf surface* is a compact complex manifold of complex dimension two which has $W = \mathbb{C}^2 - \{(0, 0)\}$ as universal covering surface. More precisely, the Hopf surface M_t is defined by $M_t = W \mid G_t$ where $G_t = \{g^m \mid m \in \mathbb{Z}\}$ and $g : (z_1, z_2) \to (\alpha z_1 + t z_2, \alpha z_2)$, that is, $\begin{pmatrix} z_1 \\ z_2 \end{pmatrix} \to \begin{pmatrix} \alpha & t \\ 0 & \alpha \end{pmatrix}\begin{pmatrix} z_1 \\ z_2 \end{pmatrix}$, where $0 < |\alpha| < 1$ and $t \in \mathbb{C}$. Then M_t is a compact complex manifold.

LEMMA 4.1. $\{M_t \mid t \in \mathbb{C}\}$ is a complex analytic family.

Proof. $= \{M_t \mid t \in \mathbb{C}\} = \mathbb{C} \times W/\Gamma$, where $\Gamma = \{\gamma^m \mid m \in \mathbb{Z}\}$, and

$$\gamma \begin{pmatrix} t \\ z_1 \\ z_2 \end{pmatrix} = \begin{pmatrix} 1 & 0 & 0 \\ 0 & \gamma & t \\ 0 & 0 & \gamma \end{pmatrix} \begin{pmatrix} t \\ z_1 \\ z_2 \end{pmatrix}. \qquad \text{Q.E.D.}$$

We claim

(1) $M_t = M_1$ (complex analytically) for $t \neq 0$.
(2) $M_0 \neq M_1$.

Proof of (1).   We make the following change of coordinates:

$$\begin{pmatrix} z_1 \\ z_2 \end{pmatrix} \rightarrow \begin{pmatrix} 1/t & 0 \\ 0 & 1 \end{pmatrix} \begin{pmatrix} z_1 \\ z_2 \end{pmatrix} = \begin{pmatrix} z_1/t \\ z_2 \end{pmatrix}.$$

Then the equation

$$\begin{pmatrix} t & 0 \\ 0 & 1 \end{pmatrix} \begin{pmatrix} \alpha & 1 \\ 0 & \alpha \end{pmatrix} \begin{pmatrix} 1/t & 0 \\ 0 & 1 \end{pmatrix} = \begin{pmatrix} \alpha & t \\ 0 & \alpha \end{pmatrix}$$

implies that $M_1 = M_t$ when $t \neq 0$.

Proof of (2).   First we prove a special case of Hartog's lemma.

LEMMA 4.2.   Any holomorphic function defined on $W = \mathbb{C}^2 - \{(0, 0)\}$ can be extended to a unique holomorphic function on $\mathbb{C}^2$.

Proof.   Let $f(z_1, z_2)$ be the function on W. Pick a number $r > 0$, and define the function

$$F(z_1, z_2) = \frac{1}{2\pi i} \oint_{|w| = r} \frac{f(w, z_2)}{w - z_1} \, dw,$$

for $|z_1| < r$ and z_2 arbitrary. Then $F(z_1, z_2)$ is an analytic function in its cylinder of definition which is a neighborhood of $(0, 0)$. If we can prove $f = F$ where both are defined, we will be finished. We know that $f(w, z_2)$ is holomorphic if $z_2 \neq 0$. So Cauchy's theorem gives

$$F(z_1, z_2) = f(z_1, z_2) \qquad \text{for } |z_1| < r, z_2 \neq 0.$$

Fix $z_1, 0 < |z_1| < r$. Then $F(z_1, z_2) = f(z_1, z_2)$ for $z_2 \neq 0$. Both are analytic in z_2; therefore,

$$F(z_1, 0) = f(z_1, 0).$$

Hence they agree where defined, proving the lemma.

Now let us suppose $M_t = M_0$, $t \neq 0$. Then there is a biholomorphic map $f : M_t \rightarrow M_0$. W is the universal covering manifold of M_t and M_0, so f induces a map $f: W \to W$ which is biholomorphic, such that

$$
\begin{array}{ccc}
W & \xrightarrow{\ f\ } & W \\
{\scriptstyle G_t}\downarrow & & \downarrow{\scriptstyle G_0} \\
M_t & \xrightarrow{\ f\ } & M_0
\end{array}
$$

commutes.

It follows that $G_t = f^{-1} G_0 f$. Hence for generator g_t of G_t,

$$
g_t = f^{-1} g_0^{\pm 1} f. \tag{9}
$$

Write the map f in coordinates as

$$
f(z_1, z_2) = [f_1(z_1, z_2), f_2(z_1, z_2)].
$$

Then by Hartog's lemma extend $f_\lambda(z_1, z_2)$ to a holomorphic function F_λ (z_1, z_2) on $\mathbb{C}^2$. Then F maps $\mathbb{C}^2$ into $\mathbb{C}^2$ $[F = (F_1, F_2)]$, and $F(0) = 0$. For if not, extend f^{-1} to $\hat{F}$ which satisfies $\hat{F}[F(z)] = z$ on W and by continuity, $\hat{F}[F(0)] = 0$. But if $F(0) \neq 0$, $\hat{F}[F(0)] = f^{-1}[F(0)] \neq 0$. This contradiction gives the result. Now expand F_λ,

$$
F_\lambda(z_1, z_2) = F_{\lambda_1} z_1 + F_{\lambda_2} z_2 + F_{\lambda_3} z_1^2 + F_{\lambda_4} z_1 z_2 + \cdots .
$$

We know that $f[g_t(z)] = g_0^{+1}[f(z)]$ so

$$
F[g_t(z)] = \begin{pmatrix} \alpha & 0 \\ 0 & \alpha \end{pmatrix}^{\pm 1} F(z).
$$

Rewriting this gives

$$
F_1(\alpha z_1 + t z_2, \alpha z_2) = \alpha^{\pm 1} F_1(z_1, z_2),
$$
$$
F_2(\alpha z_1 + t z_2, \alpha z_2) = \alpha^{\pm 1} F_1(z_1, z_2).
$$

Expanding these and taking the linear terms yields

$$
\begin{pmatrix} F_{11} & F_{12} \\ F_{21} & F_{22} \end{pmatrix} \begin{pmatrix} \alpha & t \\ 0 & \alpha \end{pmatrix} = \begin{pmatrix} \alpha & 0 \\ 0 & \alpha \end{pmatrix}^{\pm 1} \begin{pmatrix} F_{11} & F_{12} \\ F_{21} & F_{22} \end{pmatrix}.
$$

This can only happen when $t = 0$. Hence $M_1 \neq M_0$.   Q.E.D.

EXAMPLE 4.   Ruled Surfaces (examples of surgery)   Our ruled surfaces will be $\mathbb{P}^1$ bundles over $\mathbb{P}^1$. Let $\mathbb{P}^1 = \{\zeta \mid \zeta \in \mathbb{C} \cup \{\infty\}\}$ (nonhomogeneous coordinates). $M^{(m)} = U_1 \times \mathbb{P}^1 \cup U_2 \times \mathbb{P}^1$ where $U_1 \cup U_2 = \mathbb{P}^1$, $U_1 = \mathbb{C}$, $U_2 = \mathbb{P}^1 - \{0\}$, and identification takes place as follows (recall Section 3): Let $(z_1, \zeta_1) \in U_1 \times \mathbb{P}^1$, $(z_2, \xi_2) \in U_2 \times \mathbb{P}^1$. Then

$$
(z_1, \zeta_1) \leftrightarrow (z_2, \zeta_2) \qquad \text{if } \zeta_1 = z_2^m \zeta_2,\ z_1 = 1/z_2.
$$

REMARK. $M^{(m)} \neq M^{(\ell)}$ for $m \neq \ell$ (not to be proved now).

THEOREM 4.2. $M^{(\ell)}$ is a deformation of $M^{(m)}$ if $m - \ell \equiv 0$ (mod 2). Assume that $m > \ell$. Then there is a complex analytic family $\{M_t \mid t \in C\}$ such that $M_0 = M^{(m)}$ and $M_t = M^{(\ell)}$ for $t \neq 0$.

Proof.   Define M_t as follows: $M_t = U_1 \times \mathbb{P}^1 \cup U_2 \times \mathbb{P}^1$ where (z_1, ζ_1) $\leftrightarrow (z_2, \zeta_2)$ if $z_1 = 1/z_2$, $\zeta_1 = z_0^m \zeta_2 + t z_2^k$ where $k = \frac{1}{2}(m - \ell)$. Then it is easy to see that $\{M_t, t \in C\}$ is a complex analytic family and that $M_0 = M^{(m)}$. Suppose $t \neq 0$. Introduce new coordinates on the first $\mathbb{P}_1$ by

$$\zeta_1' = \frac{z_1^k \zeta_1 - t}{t \zeta_1} \quad \text{(linear fractional transformation)}.$$

On the second $\mathbb{P}^1$,

$$\zeta_2' = \frac{\zeta_2}{t z_2^{m-k} \zeta_2 + t^2}.$$

Then, using $z_1 z_2 = 1$, and $\zeta_1 = z_2^m \zeta_2 + t z_2^k$, we get

$$\zeta_1' = z_2^{m-2k} \zeta_2'.$$

Hence, in the new coordinates, $z_1 z_2 = 1$, $\zeta_1' = z_2^\ell \zeta_2'$; so

$$M_t = M^{(\ell)} \qquad \text{for} \quad t \neq 0. \qquad \text{Q.E.D.}$$

PROBLEM.   Find a pair of complex analytic families $\{M_t \mid |t| < 1\}, \{N_t \mid |t| < 1\}$ such that

> (a) $M_0 \neq N_0$,   (not complex analytically
> (b) $M_t = N_0$ for $t \neq 0$,   homeomorphic)
> (c) $N_t = M_0$ for $t \neq 0$.

There are no known examples of this type.

[2]

Sheaves and Cohomology

I.　Germs of Functions

Let M be a complex (or differentiable) manifold. A *local holomorphic (differentiable) function* is a holomorphic (differentiable) function defined on an open subset $U \subseteq M$. We write $D(f)$ for the domain of f. Let $p \in M$ and suppose given local functions f, g such that $D(f) \cap D(g) \ni p$. We say that f and g are equivalent at p if $f(z) = g(z)$ for $z \in W \subseteq D(f) \cap D(g)$, W a neighborhood of p. By a *germ of a function at p* we mean an equivalence class of local functions at p. Denote by f_p the germ of f at p, $\mathcal{O}_p$ the set of germs of all holomorphic functions at p, and $\mathcal{D}_p$ the set of germs of all differentiable functions at p. The definitions

$$\alpha f_p + \beta g_p = (\alpha f + \beta g)_p \qquad \alpha, \beta \in \mathbb{C},$$

$$f_p \cdot g_p = (fg)_p,$$

are well defined, hence, $\mathcal{O}_p$, $\mathcal{D}_p$ become linear spaces over $\mathbb{C}$. We also define,

$$\mathcal{O} = \bigcup_{p \in M} \mathcal{O}_p, \quad \mathcal{D} = \bigcup_{p \in M} \mathcal{D}_p.$$

We put a topology on $\mathcal{O}$ and $\mathcal{D}$ as follows: Take any $\varphi \in \mathcal{O}$ (or $\mathcal{D}$); then $\varphi \in \mathcal{O}_p$ (or $\mathcal{D}_p$) for some p. Take any holomorphic (differentiable) f with $f_p = \varphi$ and define a neighborhood of φ as follows:

$$\mathcal{U}(\varphi; f, U) = \{f_q \mid q \in U\},$$

where $p \in U \subseteq M$, U is an open set in $D(f)$. It is easy to see that the system of neighborhoods $\mathcal{U}(\varphi; f, U)$ defines a topology on $\mathcal{O}$ (or $\mathcal{D}$).

EXAMPLE.　$\mathcal{O}$ on the complex plane $\mathbb{C}$. Let $p \in \mathbb{C}$. Then if f and g are holmorphic at p we have expansions valid in some neighborhood of p,

$$f(z) = \sum_{k=0}^{\infty} f_k(z - p)^k, \quad g(z) = \sum_{k=0}^{\infty} g_k(z - p)^k,$$

so f and g are equivalent at p if and only if $f_k = g_k$ for all k. Hence, the germ at p is represented by a convergent power series; $\mathcal{O}_p =$ ring of convergent power series. And an element $\varphi \in \mathcal{O}_p$ can be represented by $\varphi = f_p = \{p; f_0, f_1, \cdots\}$ where $\overline{\lim}_{k \to \infty} |f_k|^{1/k} < +\infty$ and the radius of convergence is $r(\varphi) = 1/\overline{\lim}$.

27

We define

$$\mathcal{U}(\varphi; \varepsilon) = \{\psi \mid \psi = f_q, |q - p| < \varepsilon \text{ where } 0 < \varepsilon < r(\varphi)\}.$$

In terms of our representation we calculate

$$f(z) = \sum_{k=0}^{\infty} f_k(z - p)^k = \sum_{m=0}^{\infty} f_m(z - q + q - p)^m$$

$$= \sum_{k=0}^{\infty} (z - q)^k \sum_{m=k}^{\infty} \binom{m}{k} f_k(q - p)^{m-k}.$$

Hence

$$\mathcal{U}(\varphi; \varepsilon) = \left\{ \psi \mid \psi = (q; g_0, \cdots, g_k, \cdots), |(q - p)| < \varepsilon \right.$$

$$g_k = \left. \sum_{m=k}^{\infty} \binom{m}{k} f_m(q - p)^{m-k} \right\}.$$

We note that $\psi \in \mathcal{U}(\varphi; \varepsilon)$ means that ψ is a direct analytic continuation of φ.

The case of $\mathcal{D}$ on $\mathbb{R}$ is not so simple. If $\varphi = f_p$ where f is a C^∞ function at p,

$$f(x) = \sum_{k=0}^{m} f_k(x - p)^k + 0(x - p)^m.$$

But f is *not* determined by the f_k's since there exist C^∞ functions f which are not identically zero, but which have all derivatives zero at some point.

Define $\bar{\omega}: \mathcal{O}$ (or $\mathcal{D}) \to M$ by $\bar{\omega}(\mathcal{O}_p) = p$.

PROPOSITION 1.1. (1) $\bar{\omega}$ is a local homeomorphism (that is, there exists $\mathcal{U}$ such that $\bar{\omega}: \mathcal{U}(\varphi; f, U) \to U$ is a homeomorphism).

(2) $\bar{\omega}^{-1}(p) = \mathcal{O}_p$ (or $\mathcal{D}_p$) (obvious).

(3) The module operations on $\bar{\omega}^{-1}(p)$ are continuous (that is, $\alpha\varphi + \beta\psi$ depends continuously on φ, ψ).

Proof. (1) $\mathcal{U}(\varphi; f, U) = \{f_q \mid q \in U\}$ and $\bar{\omega}: f_q \to q$ is certainly $1 - 1$. It is obvious that $\bar{\omega}$ is continuous. To show that $\bar{\omega}^{-1}$ is continuous, let $\mathcal{U}(\omega; g, V)$ be a neighborhood of $\psi = f_q$. We want to find a neighborhood W of q so that $f_w = \bar{\omega}^{-1}(w) \in \mathcal{U}(\psi; g, V)$ for $w \in W$. We know that $g_q = \psi = f_q$, so f and g are equivalent at q. Hence, $f = g$ in some neighborhood N of q. Let $W = N \cap V$. Then $f_w = g_w$ on W, so $f_w \in \mathcal{U}(\psi; g, V)$ for $w \in W$. This proves that the $\bar{\omega}^{-1}$ is continuous.

(3) Let $\varphi = f_p$, $\psi = g_p$. Then $\alpha\varphi + \beta\psi = (\alpha f + \beta g)_p$. Let $\mathcal{U}(\alpha\varphi + \beta\psi; h, U)$ be a neighborhood of $\alpha\varphi + \beta\psi$. Then $\alpha\varphi + \beta\psi = h_p = (\alpha f + \beta g)_p$ so

$h = \alpha f + \beta g$ in some neighborhood $V \subseteq U$ of p. Then if $\sigma \in \mathcal{U}(\varphi; f, V)$, $\tau \in \mathcal{U}(\psi; g, V)$, we have

$$\begin{aligned}
\alpha\sigma + \beta\tau &= \alpha f_q + \beta g_q \\
&= (\alpha f + \beta g)_q \\
&= h_q \in \mathcal{U}(\alpha\varphi + \beta\psi; h, V).
\end{aligned}$$

Since $\mathcal{U}(\alpha\varphi + \beta\psi; h, V) \subseteq \mathcal{U}(\alpha\varphi + \beta\psi; h, U)$ we are done.   Q.E.D.

We now give a formal definition. Let X be a paracompact Hausdorff space.

DEFINITION 1.1.   A *sheaf* $\mathcal{S}$ over X is a topological space with a map $\bar{\omega}$: $\mathcal{S} \to X$ onto X such that

(1) $\bar{\omega}$ is a local homeomorphism [that is, each point $s \in \mathcal{S}$ has a neighborhood $\mathcal{U}$ such that $\bar{\omega}: \mathcal{U} \to \bar{\omega}(\mathcal{U}) \subset X$ is a homeomorphism onto an open neighborhood of $\bar{\omega}(s)$].
(2) $\bar{\omega}^{-1}(x), x \in X$ is an R-module where $R = \mathbb{Z}, \mathbb{R}, \mathbb{C}$, or principal ideal ring.
(3)   The module operation $(s, t) \to \alpha s + \beta t$ is continuous on $\bar{\omega}^{-1}(x)$ where $\alpha, \beta \in R$.

(The reader can easily generalize this definition, but for our purposes it suffices.) The set $\mathcal{S}_x = \bar{\omega}^{-1}(x)$ is called the *stalk* of $\mathcal{S}$ over x.

EXAMPLES.   (of sheaves)

(1) $\mathcal{O}$ on a complex manifold.
(2) $\mathcal{D}$ on a differentiable manifold.
(3)   The sheaf over X of germs of continuous ($\mathbb{R}$ or $\mathbb{C}$ valued) functions.
(4)   The sheaf over X of germs of constant functions.

In Example (4) $\mathcal{S} = X \times \mathbb{C}$ with the following topology: Let $s = (x, z)$; then $\mathcal{U}(s) = \{(y, z) \mid y \in U, z \text{ fixed}\}$. If $r \to f(r)$ is a continuous map into $\mathcal{S}$ of $I = \{r \mid a < r < b\}$, then $f(I) = \{(y, z) \mid z \text{ fixed and } y = \bar{\omega}(f(r)) r \in I\}$. In other words we give $X \times \mathbb{C}$ the product topology where X has its given topology and $\mathbb{C}$ has the discrete topology.

DEFINITION 1.2.   Let U be a subset (usually open) of X. By a *section* σ of $\mathcal{S}$ over U we mean a continuous map $x \to \sigma(x)$ such that $\bar{\omega}\,\sigma(x) = x$. Suppose $X = M$, a complex (or differentiable) manifold; and suppose $\mathcal{S} = \mathcal{O}$ (or $\mathcal{D}$). If $f(z)$ is a holomorphic (or differentiable) function on U, then $\sigma: p \to f_p, p \in U$ is a section.

PROPOSITION 1.2.   Let $\sigma: U \to \mathscr{S}$ be a section ($\mathscr{S}$ as above). Then σ determines a holomorphic (or differentiable function) $f = f(z)$ on U such that $\sigma(p) = f_p$.

Proof. $\sigma(p) \in \mathcal{O}_p$ (or $\mathcal{D}_p$). Hence there is a holomorphic (or differentiable) $g(z)$ defined on some neighborhood of p so that $\sigma(p) = g_p$. Since g depends on p we write, $g(z) = g^{(p)}(z)$. Define f as follows:

$$f(p) = g^{(p)}(p).$$

Then f is obviously well defined. Then

(1) $f(p)$ is a holomorphic (differentiable) function on U.

Proof.   Take W_a neighborhood of p, $W \subseteq U$. Let $\mathscr{U} = \mathscr{U}[\sigma(p); g^{(p)}, W]$ $= \{(g^{(p)})_q \mid q \in W\}$. Since σ is continuous, for any small neighborhood N of p, $N \subseteq W$, we have $\sigma(N) \subseteq \mathscr{U}$. Hence $\sigma(q) = (g^{(p)})_q$. But we also know $\sigma(q) = (g^{(q)})_q$. Thus, $(g^{(q)})_q = (g^{(p)})_q$, and $g^{(q)}(z) = g^{(p)}(z)$ for z in a small neighborhood V of q, $V \subset N$. But $f(q) = g^{(q)}(q) = g^{(p)}(q)$ for $q \in V$. So $f(z) = g^{(p)}(z)$ for $z \in V$ and $g^{(p)}$ holomorphic (or differentiable) in V implies that f is also.

(2)   By definition $\sigma(p) = (g^{(p)})_p = f_p$ for each $p \in U$.   Q.E.D.

Hence we have the maps:

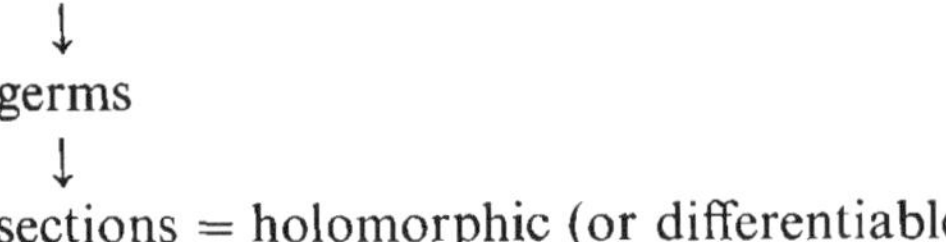

$\Gamma(U, \mathscr{S})$ will denote the R-module consisting of all sections of $\mathscr{S}$ over U. We remark that $\Gamma(U, \mathcal{O})$ are all holomorphic functions over U and $\Gamma(U, \mathcal{D})$ are all differentiable functions over U. Let $\{U_\lambda \mid 1 \leq \lambda \leq n\}$ be a finite family of open sets in X such that $\cap U_\lambda \neq \phi$. Let $\sigma_\lambda \in \Gamma(U_\lambda, \mathscr{S})$ and $\alpha_\lambda \in R$. Then $\sum \alpha_\lambda \sigma_\lambda \in \Gamma(U, \mathscr{S})$ where $U = \cap U_\lambda$. Let W be an open set and $\sigma \in \Gamma(U, \mathscr{S})$ for some open set U. Then $x \to \sigma(x)$, $x \in W \cap U$ defines a section of $\Gamma(W \cap U, \mathscr{S})$. We denote this section by $r_w \sigma$ and call it the restriction of σ to $W \cap U$.

2.   Cohomology Groups

Let X be a Hausdorff paracompact space and let $\mathscr{S}$ be a sheaf over X. Fix a locally finite covering $\mathscr{U} = \{U_j\}$ of X. A *0-cochain* C^0 on X is a set $C^0 = \{\sigma_j\}$ of sections $\sigma_j \in \Gamma(U_j,)$. A *1-cochain* $C^1 = \{\sigma_{jk}\}$ is a set of sections

$\sigma_{jk} \in \Gamma(U_j \cap U_k, \mathscr{S})$ such that $\sigma_{jk} = -\sigma_{kj}$ (skew-symmetric). A *q-cochain* $C^q = \{\sigma_{j_0 \cdots j_k}\}$ is a set of sections $\sigma_{j_0 \cdots j_k} \in \Gamma(U_{j_0} \cap \cdots \cap U_{j_k}, \mathscr{S})$ which are skew-symmetric in the indices $j_0 \cdots j_k$. Let $C^q(\mathscr{U})$ be the R-module of all q-cochains. We define a map $C^q(\mathscr{U}) \xrightarrow{\delta} C^{q+1}(\mathscr{U})$, the *coboundary map* as follows: For 0-cochains, $\delta C^0 = \{\tau_{jk}\} = \{\sigma_k - \sigma_j\}$ where $C^0 = \{\sigma_k\}$; for 1-cochains $C^1 = \{\sigma_{jk}\}$, $\delta C^1 = \{\tau_{jk\ell}\}$ where $\tau_{jk\ell} = \sigma_{k\ell} - \sigma_{j\ell} + \sigma_{jk} = \sigma_{jk} + \sigma_{k\ell} + \sigma_{\ell j}$. In general, $\delta C^q = \{\tau_{j_0 \cdots j_{q+1}}\}$ if $C^q = \{\sigma_{j_0 \cdots j_q}\}$, where

$$
\begin{aligned}
\tau_{j_0 \cdots j_{q+1}} &= \sigma_{j_1 \cdots j_{q+1}} - \sigma_{j_0 j_2 \cdots j_{q+1}} + \cdots \\
&\quad + (-1)^{q+1} \sigma_{j_0 \cdots j_q} \\
&= \sum (-1)^k \sigma_{j_0 \cdots \hat{j_k} \cdots j_{q+1}},
\end{aligned}
\tag{1}
$$

where $\wedge$ means "omit."

We denote the *q-cocycles* by

$$
Z^q(\mathscr{U}) = \{C^q \mid \delta C^q = 0\}.
$$

The *q-cohomology group* (with respect to $\mathscr{U}$) is

$$
H^q(\mathscr{U}) = H^q(\mathscr{U}, \mathscr{S}) = Z^q(\mathscr{U}) / \delta C^{q-1}(\mathscr{U}).
\tag{2}
$$

We should remark that δC is always skew-symmetric and $\delta\delta = 0$ so that $\delta C^{q-1}(\mathscr{U}) \subseteq Z^q(\mathscr{U})$ and Equation (2) makes sense. The *qth cohomology group of X with coefficients in the sheaf $\mathscr{S}$* is defined to be

$$
H^q(X, \mathscr{S}) = \lim_{\mathscr{U}} H^q(\mathscr{U}, \mathscr{S}).
$$

This limiting process will now be explained. We say that the open covering $\mathscr{V} = \{V_\lambda\}_{\lambda \in \Lambda}$ of X is a *refinement* of $\mathscr{U} = \{U_j\}_{j \in J}$ if there is a map $s: \Lambda \to J$ such that $V_\lambda \subset U_{s(\lambda)} = U_{j(\lambda)}$, where we set $j(\lambda) = s(\lambda)$. We define a homomorphism

$$
\Pi^{\mathscr{U}}_{\mathscr{V}} : C^q(\mathscr{U}) \to C^q(\mathscr{V}),
$$

$$
\Pi^{\mathscr{U}}_{\mathscr{V}} : \{\sigma_{j_0 \cdots j_q}\} \to \{\tau_{\lambda_0 \cdots \lambda_q}\},
$$

where

$$
\tau_{\lambda_0 \cdots \lambda_q} = r_{V_{\lambda_0} \cap \cdots \cap V_{\lambda_q}}[\sigma_{s(\lambda_0) \cdots \lambda(\lambda_q)}].
\tag{3}
$$

It is easy to check that

$$
\delta \Pi^{\mathscr{U}}_{\mathscr{V}} = \Pi^{\mathscr{U}}_{\mathscr{V}} \delta,
\tag{4}
$$

so that $\Pi^{\mathscr{U}}_{\mathscr{V}}$ maps $Z^q(\mathscr{U})$ into $Z^q(\mathscr{V})$ and $\delta C^{q-1}(\mathscr{U})$ into $\delta C^{q-1}(\mathscr{V})$. Hence $\Pi^{\mathscr{U}}_{\mathscr{V}}$ induces a homomorphian $\Pi^{\mathscr{U}}_{\mathscr{V}}: H^q(\mathscr{U}) \to H^q(\mathscr{V})$.

LEMMA 2.1. $\Pi^{\mathscr{U}}_{\mathscr{V}}: H^q(\mathscr{U}) \to H^q(\mathscr{V})$ is independent of the choice of map $s: \Lambda \to J$ in the definition of refinement.

Proof. First some notation: fix indices $\alpha_0, \cdots, \alpha_q \in \Lambda$. Let

$$V = V_{\alpha_0} \cap \cdots \cap V_{\alpha_q}, \quad V^\ell = V_{\alpha_0} \cap \cdots \cap \widehat{V_{\alpha_\ell}} \cap \cdots \cap V_{\alpha_q},$$

$$\widehat{U^{j\ell}} = U_{f(\alpha_0)} \cap \cdots \cap U_{f(\alpha_j)} \cap U_{g(\alpha_j)} \cap \cdots \widehat{U_{g(\alpha_\ell)}} \cap \cdots \cap U_{g(\alpha_q)},$$

and

$$U^j = U_{f(\alpha_1)} \cap \cdots \cap U_{f(\alpha_j)} \cap U_{g(\alpha_j)} \cap \cdots \cap U_{g(\alpha_q)},$$

where $f, g : \Lambda \to J$ are two refining maps. Define a function $(k\sigma)_{\lambda_1 \cdots \lambda_q}$ by

$$(k\sigma)_{\lambda_1 \cdots \lambda_q} = \sum_{p=0}^{q} (-1)^{p-1} r_V \circ \sigma_{f(\lambda_1) \cdots f(\lambda_p)g(\lambda_p) \cdots g(\lambda_q)} \tag{5}$$

Let us call the maps $\Pi^{\mathcal{U}}_{\mathcal{V}}$, defined by f and g, f^*, and g^*. We claim that the following equation holds:

$$[(\delta k + k\delta)]_{\alpha_0 \cdots \alpha_q} = (g^*\sigma - f^*\sigma)_{\alpha_0 \cdots \alpha_q}. \tag{6}$$

The function $k\sigma$ is not necessarily skew-symmetric in its indices; so we skew-symmetrize

$$\tau'_{\lambda_1 \cdots \lambda_q} = (k'\tau)_{\lambda_1 \cdots \lambda_q} = \frac{1}{q!} \sum \operatorname{sgn} \begin{pmatrix} \lambda_1 & \cdots & \lambda_q \\ \mu_1 & \cdots & \mu_q \end{pmatrix} \tau_{\mu_1 \cdots \mu_q}.$$

Next we use (6) to see that

$$[(\delta k' + k'\delta)\sigma]_{\alpha_0 \cdots \alpha_q} = (g^*\sigma - f^*\sigma)_{\alpha_0 \cdots \alpha_q}.$$

Hence, if $\delta\sigma = 0$, $\delta k'\sigma = g^*\sigma - f^*\sigma \in \delta C^{q-1}(\mathcal{V})$. Hence, f^* and g^* induce the same map, $H^q(\mathcal{U}) \to H^q(\mathcal{V})$. Therefore we prove (6). The reader can easily check the following calculations:

$$(\delta k\sigma)_{\alpha_0 \cdots \alpha_q} = \sum_{l=0}^{q} (-1)^\ell r_V (k\sigma)_{\alpha_0 \cdots \widehat{\alpha_l} \cdots \alpha_q}$$

$$= \sum_{\ell=0}^{q} (-1)^\ell r_V \left[\sum_{j=0}^{\ell-1} (-1)^j r_{V\ell}\, \sigma_{f(\alpha_0) \cdots f(\alpha_j g(\alpha_j) \cdots \widehat{g(\alpha_l)} \cdots g(\alpha_q)} \right.$$

$$\left. + \sum_{j=\ell+1}^{q} (-1)^{j-1} r_{V\ell}\, \sigma_{f(\alpha_0) \cdots \widehat{f(\alpha_l)}f(\alpha_j)g(\alpha_j) \cdots g(\alpha_q)} \right]$$

$$(\delta k\sigma)_{\alpha_0 \cdots \alpha_q} = \sum_{j<\ell} (-1)^{\ell+j} r_V\, \sigma_{f(\alpha_0) \cdots f(\alpha_j)g(\alpha_j) \cdots \widehat{g(\alpha_l)} \cdots g(\alpha_q)}$$

$$+ \sum_{j>\ell} (-1)^{\ell+j+1} r_V\, \sigma_{f(\alpha_0) \cdots f(\alpha_l) \cdots f(\alpha_j)g(\alpha_j) \cdots g(\alpha_q)}. \tag{7}$$

Similarly,

$$(k\delta\sigma)_{\alpha_0 \cdots \alpha_q} = \sum_{\ell \le j} (-1)^{j+\ell} r_V\, \sigma_{f(\alpha_0) \cdots \widehat{f(\alpha_l)} \cdots f(\alpha_j)g(\alpha_j) \cdots g(\alpha_q)}$$

$$+ \sum_{\ell \ge j} (-1)^{j+\ell+1} r_V\, \sigma_{f(\alpha_0) \cdots f(\alpha_j)g(\alpha_j) \cdots \widehat{g(\alpha_l)} \cdots g(\alpha_q)}. \tag{8}$$

Equations (7) and (8) give

$$[(\delta k + k\delta)\sigma]_{\alpha_0 \cdots \alpha_q} = \sum_{j=0}^{q} r_V \, \sigma_{f(\alpha_0) \cdots \widehat{f(\alpha_0)}g(\alpha_j) \cdots g(\alpha_q)}$$

$$- \sum_{j=0}^{q} r_V \, \sigma_{f(\alpha_0) \cdots f(\alpha_j)g(\alpha_j) \cdots g(\alpha_q)}$$

$$= r_V \, \sigma_{g(\alpha_0) \cdots g(\alpha_q)} - r_V \, \sigma_{f(\alpha_0) \cdots f(\alpha_q)}, \qquad (9)$$

proving Equation (6).   Q.E.D.

Knowing that the map $\Pi_{\mathscr{V}}^{\mathscr{U}}$ depends only on $\mathscr{U}$ and $\mathscr{V}$, we proceed to the definition of the limit. We write $\mathscr{U} < \mathscr{W}$ if $\mathscr{W}$ is a locally-finite refinement of $\mathscr{U}$. Then $<$ is a partial order and given $\mathscr{U}$, $\mathscr{V}$ there is $\mathscr{W}$ so that $\mathscr{U} < \mathscr{W}$ and $\mathscr{V} < \mathscr{W}$. Hence the set of all locally finite coverings of X forms a directed set with respect to $<$, and the following equations can be verified (using Lemma 2.1):

$$\Pi_{\mathscr{U}}^{\mathscr{U}} = id,$$

$$\Pi_{\mathscr{W}}^{\mathscr{U}} = \Pi_{\mathscr{W}}^{\mathscr{V}} \circ \Pi_{\mathscr{V}}^{\mathscr{U}}, \qquad \text{if} \quad \mathscr{U} < \mathscr{V} < \mathscr{W}.$$

DEFINITION 2.1. $H^q(X, \mathscr{S}) = \lim_{\mathscr{U}} H^q(\mathscr{U}, \mathscr{S}).$

REMARK.   We recall the definition of the limit $\lim_{\mathscr{U}}$. We say that $g, h \in H^q$ $(\mathscr{U}, \mathscr{S})$ are equivalent if there exists $\mathscr{W} > \mathscr{U}$ such that $\Pi_{\mathscr{W}}^{\mathscr{U}} g = \Pi_{\mathscr{W}}^{\mathscr{U}} h$. Denote the equivalence class of g by $\bar{g}$. Let

$$\bar{H}^q(\mathscr{U}, \mathscr{S}) = \{\bar{g} \mid g \in H^q(\mathscr{U}, \mathscr{S})\}.$$

The map $g \to \bar{g}$ defines a homomorphism $\Pi^{\mathscr{U}}$,

$$\Pi_{\mathscr{V}}^{\mathscr{U}} : H^q(\mathscr{U}, \mathscr{S}) \to \bar{H}^q(\mathscr{V}, \mathscr{S}),$$

and $\Pi_{\mathscr{V}}^{\mathscr{U}}$ induces a homomorphism $\bar{\Pi}_{\mathscr{V}}^{\mathscr{U}}$,

$$\bar{\Pi}_{\mathscr{V}}^{\mathscr{U}} : \bar{H}^q(\mathscr{U}, \mathscr{S}) \to \bar{H}^q(\mathscr{V}, \mathscr{S}).$$

LEMMA 2.2. $\bar{\Pi}_{\mathscr{V}}^{\mathscr{U}}$ is injective.

Proof. $\bar{\Pi}_{\mathscr{V}}^{\mathscr{U}}\bar{g} = 0$ if and only if $\Pi_{\mathscr{W}}^{\mathscr{V}} \circ \Pi_{\mathscr{V}}^{\mathscr{U}} g = 0$ for some W. So $\Pi_{\mathscr{W}}^{\mathscr{U}} g = 0$ and $\bar{g} = 0$.   Q.E.D.

Hence, identifying $\bar{H}^q(\mathscr{U}, \mathscr{S})$ with $\bar{\Pi}_{\mathscr{V}}^{\mathscr{U}} \bar{H}^q(\mathscr{U}, \mathscr{S})$, we may consider $\bar{H}^q(\mathscr{U}, \mathscr{S}) \subset \bar{H}^q(\mathscr{V}, \mathscr{S})$ provided that $\mathscr{U} < \mathscr{V}$. Then by definition,

$$H^q(X, \mathscr{S}) = \bigcup_{\mathscr{U}} \bar{H}^q(\mathscr{U}, \mathscr{S}),$$

and $\Pi^{\mathscr{U}} : H^q(\mathscr{U}, \mathscr{S}) \to \bar{H}^q(\mathscr{U}, \mathscr{S}) \subseteq H^q(X, \mathscr{S})$ is a homomorphism of $H^q(\mathscr{U}, \mathscr{S})$ into $H^q(X, \mathscr{S})$.

PROPOSITION 2.1. $H^0(X, \mathscr{S}) = \Gamma(X, \mathscr{S})$.

Proof. By definition $C^{-1} = 0$ so $H^0(\mathscr{U}, \mathscr{S}) = Z^0(\mathscr{U}, \mathscr{S})$.

$$Z^0(\mathscr{U}, \mathscr{S}) = [\sigma \mid \sigma = \{\sigma_j\}, \sigma_j \in \Gamma(U_j, \mathscr{S}), \delta\sigma = 0].$$

But $\delta\sigma = 0$ means $\sigma_j(z) - \sigma_k(z) = 0$ on $U_j \cap U_k$. Hence $\sigma(z) \in \Gamma(X, \mathscr{S})$, defined by $\sigma(z) = \sigma_j(z)$ when $z \in U_j$, is meaningful. This proves $H^0(\mathscr{U}, \mathscr{S}) = \Gamma(X, \mathscr{S})$ and implies $H^0(X, \mathscr{S}) = \Gamma(X, \mathscr{S})$.　Q.E.D.

PROPOSITION 2.2. $H^{\mathscr{U}} : H^1(\mathscr{U}, \mathscr{S}) \to H^1(X, \mathscr{S})$ is injective.

COROLLARY. $H^1(X, \mathscr{S}) = \bigcup_{\mathscr{U}} H^1(\mathscr{U}, \mathscr{S})$.

Proof. (of the proposition). Suppose $h \in H^1(\mathscr{U}, \mathscr{S}) = Z^1(\mathscr{U})/\delta C^0(\mathscr{U})$. Then $h = \{\sigma_{jk}\}, \sigma_{jk} \in \Gamma(U_j \cap U_k, \mathscr{S})$ where $\sigma_{ij} + \sigma_{jk} + \sigma_{ki} = 0$. We want to show that $\Pi^{\mathscr{U}} h = 0$ implies $h = 0$. $\Pi^{\mathscr{U}} h = 0$ means $\bar{h} = 0$ and this is true if and only if $\Pi^{\mathscr{U}}_{\mathscr{V}} h = 0$ for some $\mathscr{V}, \mathscr{V} > \mathscr{U}$. Let $\mathscr{W} = \{W_{i\lambda} \mid W_{i\lambda} = U_i \cap V_\lambda\}$. Then $\mathscr{W}$ is a locally finite refinement of $\mathscr{V}$ and $\Pi^{\mathscr{U}}_{\mathscr{W}} h = \Pi^{\mathscr{V}}_{\mathscr{W}} \circ \Pi^{\mathscr{U}}_{\mathscr{V}} h = 0$. Also $\mathscr{W} > \mathscr{U}$ since $\mathscr{W}_{i\lambda} \subset U_i$ and we can use the map $s(i\lambda) = i$ in the definition of refinement. Then we have

$$\Pi^{\mathscr{U}}_{\mathscr{W}} h = \{\tau_{(i\lambda)(j\mu)}\}$$

where

$$\tau_{(i\lambda)(j\mu)} = \tau_{i\lambda j\mu} = r_{W_{i\lambda} \cap W_{j\mu}} \sigma_{ij}.$$

Then $\Pi^{\mathscr{U}}_{\mathscr{W}} h = 0$ implies $\{\tau_{i\lambda j\mu}\} = \delta\{\tau_{i\lambda}\}$, that is, $\tau_{i\lambda j\mu} = \tau_{j\mu} - \tau_{i\lambda}$. Since $\tau_{i\lambda i\mu} = r_{W_{i\lambda} \cap W_{i\mu}} \sigma_{ii} = 0$, we obtain $\tau_{i\mu} = \tau_{i\lambda}$ on $W_{i\lambda} \cap W_{i\mu}$. $U_i = \bigcup_\lambda W_{i\lambda}$, and $\tau_i = \tau_{i\mu}$ on $W_{i\mu}$ defines an element $\tau_i \in \Gamma(U_i, \mathscr{S})$. Then the equation $\sigma_{ij} = \tau_j - \tau_i$ implies $h = 0$.　Q.E.D.

Consequently, in order to describe an element of $H^1(X, \mathscr{S})$, it is sufficient to give an element of $H^1(\mathscr{U}, \mathscr{S})$ for some $\mathscr{U}$.

EXAMPLE. Let $M = \{(z_1, z_2) \mid |z_1| < 1, |z_2| < 1, (z_1, z_2) \neq (0, 0)\}$. Then $\dim_{\mathbb{C}} H^1(M, \mathcal{O}) = +\infty$.

Proof. Set

$$U_1 = \{(z_1, z_2) \mid (z_1, z_2) \in M, z_1 \neq 0\},$$
$$U_2 = \{(z_1, z_2) \mid (z_1, z_2) \in M, z_2 \neq 0\}.$$

In this case $M = U_1 \cup U_2$ so chose as covering $\mathcal{U} = \{U_1, U_2\}$. Then $H^1(\mathcal{U}, \mathcal{O}) = Z^1(\mathcal{U}, \mathcal{O})/\delta C^0(\mathcal{U}, \mathcal{O})$ where $Z^1(\mathcal{U}, \mathcal{O}) = \{\sigma_{12} \mid \sigma_{12} \in \Gamma(U_1 \cap U_2, \mathcal{O})\}, C^0(\mathcal{U}, \mathcal{O}) = \{\tau \mid \tau = (\tau_1, \tau_2), \tau_\mu \in \Gamma(U_\mu, \mathcal{O})\}$, and $\delta C^0(\mathcal{U}, \mathcal{O}) = \{\tau_2 - \tau_1 \mid \tau_\mu \in \Gamma(U_\mu, \mathcal{O})\}$.

We note that $U_1 \cap U_2 = \{(z_1, z_2) \mid 0 < |z_1| < 1, 0 < |z_2| < 1\}$, so we have a Laurent expansion for σ_{12}

$$\sigma_{12}(z) = \sum_{m=-\infty}^{+\infty} \sum_{n=-\infty}^{+\infty} a_{mn} z_1^m z_2^n .$$

τ_1 is holomorphic on $U_1 = \{(z_1, z_2) \mid 0 < |z_1| < 1, |z_2| < 1\}$ so $\tau_1(z) = \sum_{m=-\infty}^{+\infty} \sum_{n=0}^{\infty} b_{mn} z_1^m z_2^n$. Similarly for τ_2, $\tau_2(z) = \sum_{m=0}^{\infty} \sum_{n=-\infty}^{+\infty} c_{mn} z_1^m z_2^n$, and $\tau_2 - \tau_1 = \sum_{m \geq 0 \text{ or } n \geq 0} a_{mn} z_1^m z_2^n$. Then $H^1(\mathcal{U}) \cong \{\sigma_{12} \mid \sigma_{12} = \sum_{m=-\infty}^{-1} \sum_{n=-\infty}^{-1} a_{mn} z_1^m z_2^n\}$. Hence dim $H^1(\mathcal{U}, \mathcal{S}) = +\infty$ and since $H^1(\mathcal{U}, \mathcal{S}) \subseteq H^1(X, \mathcal{S})$, dim $H^1(X, \mathcal{S}) = +\infty$.   Q.E.D.

PROPOSITION 2.3. If $H^1(U_j, \mathcal{S}) = 0$ for all $U_j \in \mathcal{U}$, then $H^1(\mathcal{U}, \mathcal{S}) \cong H^1(X, \mathcal{S})$ where $\mathcal{U} = \{U_j\}$.

Proof. We already know that $H^1(\mathcal{U}, \mathcal{S}) \subseteq H^1(X, \mathcal{S})$. Hence we only need to show the following. Let $\mathcal{V} = \{V_\lambda\}$ be any locally finite covering. Let $\mathcal{W} = \{W_{j\lambda} \mid W_{j\lambda} = U_j \cap V_\lambda\}$. Then it suffices to show that $\Pi_{\mathcal{W}}^{\mathcal{U}} : H^1(\mathcal{U}) \to H^1(\mathcal{W})$ is surjective. Take a 1-cocycle $\{\sigma_{j\lambda k \nu}\}$ of $H^1(\mathcal{W})$ where $\sigma_{i\lambda j\mu} + \sigma_{j\mu k\nu} + \sigma_{k\nu i\lambda} = 0$. Then $\{\sigma_{i\lambda i\mu}\}$ for each fixed i is a 1-cocycle on the covering $\{W_{i\lambda}\}$ of U_i. Since $H^1(U_i, \mathcal{S}) = 0, H^1(\{W_{i\lambda}\}, \mathcal{S}) \subseteq H^1(U_i, \mathcal{S})$ gives $H^1(\{W_{i\lambda}\}, \mathcal{S}) = 0$ for each i. This implies the existence of $\tau_{i\lambda} \in \Gamma(W_{i\lambda}, \mathcal{S})$ such that $\sigma_{i\lambda i\mu} = \tau_{i\mu} - \tau_{i\lambda}$. Let τ be the 0-cochain $\{\tau_{i\lambda}\}$ on $\mathcal{W}$. Then $\{\sigma'_{i\lambda k\nu}\} = \{\sigma_{i\lambda k\nu}\} - \delta\tau$ defines a 1-cocycle on $\mathcal{W}$ which defines the same cohomology class in $H^1(\mathcal{W})$ as σ. From the definition of τ we see that $\sigma'_{i\lambda i\mu} = 0$. So $\sigma'_{i\lambda i\mu} + \sigma'_{i\mu k\nu} + \sigma^k_{\prime \nu i\lambda} = 0$ yields $\sigma'_{i\mu k\nu} = \sigma'_{i\lambda k\nu}$. Similarly, $\sigma'_{i\lambda k\nu} = \sigma'_{i\mu k\omega}$. Hence, $o'_{ik} = \sigma'_{i\lambda k\nu} = \sigma'_{i\nu k\nu}$, and $\sigma'_{ik} \in \Gamma(U_i \cap U_k, \mathcal{S})$. Now we have found σ'_{ik} so that $\Pi_{\mathcal{W}}^{\mathcal{U}}(\sigma'_{ik}) = \sigma'_{i\lambda k\nu}$, and $\{\sigma'_{i\lambda k\nu}\}$ is cohomologous to $\{\sigma_{i\lambda k\nu}\}$. Hence $\Pi_{\mathcal{W}}^{\mathcal{U}}$ is surjective.   Q.E.D.

3.   Infinitesimal Deformations

Using cohomology groups we will give an answer to the following problem: Let $\mathcal{M} = \{M_t / t \in B\}$ be a complex analytic family of compact complex manifolds M_t and let $t = (t^1, \cdots, t^n)$ be a local coordinate on B. The problem is to define $(\partial M_t / \partial t^\nu)$.

For this we define the sheaf of germs of holomorphic vector fields. Let M be a complex manifold and let W be an open subset of M. Let $\mathcal{U} = \{U_j, z_j\}$

be a covering of M with coordinates patches with coordinates $p \to z_i(p) = [z_i^1(p), \cdots, z_i^n(p)]$. A *holomorphic vector field* θ on W is given by a family of holomorphic functions $\{\theta_j^\alpha\}$ on $W \cap U_j$ where

$$\theta = \sum_{\alpha=1}^{n} \theta_j^\alpha(p) \frac{\partial}{\partial z_j^\alpha}$$

on $W \cap U_j$. These functions should behave as follows: On $W \cap U_k$,

$$\theta = \sum_{\beta=1}^{n} \theta_k^\beta(p) \frac{\partial}{\partial z_k^\beta}.$$

We want

$$\sum \theta_j^\alpha \frac{\partial}{\partial z_j^\alpha} = \sum \theta_k^\beta \frac{\partial}{\partial z_k^\beta},$$

so the transition equation

$$\theta_j^\alpha(p) = \sum_\beta \frac{\partial z_j^\alpha}{\partial z_k^\beta} \theta_k^\beta(p) \tag{1}$$

should be satisfied on $W \cap U_j \cap U_k$. Thus we have a definition of local holomorphic vector fields and we can define germs of local holomorphic vector fields. As notation we denote by Θ the sheaf over M of germs of holomorphic vector fields. (Later we shall give a formal definition of the holomorphic tangent bundle of a complex manifold.)

Next we want to define the *infinitesimal deformation* $(\partial M_t / \partial t_v)$. First we consider the case $B = \{t | \, |t| < r\} \subseteq \mathbb{C}$. $\mathcal{M}$ is a complex manifold and $\bar{\omega} : \mathcal{M} \to B$ is a holomorphic map satisfying the usual conditions

(1) $M_t = \bar{\omega}^{-1}(t)$;
(2) the rank of the Jacobian of $\bar{\omega} = 1 = \dim B$.

We can find an $\varepsilon > 0$ small enough so that $\bar{\omega}^{-1}(\Delta)$, $\Delta = \{t | \, |t| < \varepsilon\}$ looks as follows:

$$\bar{\omega}^{-1}(\Delta) = \bigcup_{j=1}^{J} \mathcal{U}_j \qquad \begin{array}{l}\text{(a union of a finite number} \\ \text{of open sets).}\end{array}$$

On each $\mathcal{U}_j$ there should be a coordinate system

$$p \to [z_j^1(p), \cdots, z_j^n(p), t(p)],$$

where $t(p) = \bar{\omega}(p)$ and such that $\mathcal{U}_j = \{p | \, |z_j^\alpha(p)| < \varepsilon_j, \, |t(p)| < \varepsilon\}$. We write $p = (z_j, t) = (z_j^1, \cdots, z_j^n, t)$. This construction is possible because rank $\bar{\omega} = 1$ These charts are holomorphically related so

$$z_j^\alpha(p) = f_{jk}^\alpha[z_k^1(p), \cdots, z_k^n(p), t(p)] = f_{jk}^\alpha(z_k, t)$$

on $U_j \cap U_k$. Let $U_{tj} = M_t \cap \mathcal{U}_j$, $|t| < \varepsilon$. Then set

$$\{(z_j^1 \cdots, z_j^n, t) | \, |z_j^\alpha| < \varepsilon_j\} = U_{tj},$$

so we can use $\{(z_j^1, \cdots, z_j^n) \mid |z_j^\alpha| < \varepsilon_j\}$ as coordinates on U_{tj}. The transformation $z_j^\alpha = f_{jk}^\alpha(z_k, t)$ depends on t. Consider $p \in \mathscr{U}_i \cap \mathscr{U}_j \cap \mathscr{U}_k$. Then $p = (z_i, t) = (z_j, t) = (z_k, t)$. So $z_i^\alpha = f_{ik}^\alpha(z_k, t) = f_{ij}^\alpha(z_j, t) = f_{ij}^\alpha[f_{jk}(z), t]$, where $f_{jk} = (f_{jk}^1, \cdots, f_{jk}^n)$. We set

$$\theta_{jk}(p, t) = \sum_{\alpha=1}^n \frac{\partial f_{jk}^\alpha(z_k, t)}{\partial t} \frac{\partial}{\partial z_j^\alpha}.$$

Obviously $\theta_{jk}(t) \in \Gamma(U_{tj} \cap U_{tk}, \Theta_t)$ where Θ_t is the sheaf of germs of holomorphic vector fields on M_t.

LEMMA 3.1. $\theta_{ik}(t) = \theta_{ij}(t) + \theta_{jk}(t)$ on $U_{tj} \cap U_{tk} \cap U_{ti}$.

Proof. Before beginning the proof we remark that $\theta_{ij} + \theta_{jk} + \theta_{ki} = 0$ and $\theta_{ij} = -\theta_{ji}$ is equivalent to $\theta_{ik} = \theta_{ij} + \theta_{jk}$. To prove the lemma we differentiate the transition equation to get

$$\frac{\partial f_{ik}^\alpha}{\partial t} = \frac{\partial f_{ij}^\alpha}{\partial t} + \sum_{\beta=1}^n \frac{\partial f_{ij}^\alpha}{\partial z_j^\beta} \frac{\partial f_{jk}^\beta}{\partial t}.$$

Then

$$\theta_{ik} = \sum_\alpha \frac{\partial f_{ik}^\alpha}{\partial t} \frac{\partial}{\partial z^\alpha} = \sum_\alpha \frac{\partial f_{ij}^\alpha}{\partial t} \frac{\partial}{\partial z_i^\alpha} + \sum_\beta \frac{\partial f_{jk}^\beta}{\partial t} \frac{\partial}{\partial z_j^\beta}$$

$$= \theta_{ij} + \theta_{jk}. \qquad \text{Q.E.D.}$$

*DEFINITION 3.1. $(dM_t/dt) = \{\theta_{ik}(p, t)\} \in H^1(M_t, \Theta_t)$. We have made several choices in this definition and we must justify them.

PROPOSITION 3.1. (dM_t/dt) is independent of the choice of local coordinate covering $\{z_j^\alpha\}$.

Proof. Let $\{\mathscr{V}_\lambda\}$ be a locally finite refinement of $\{\mathscr{U}_j\}$ such that $(\zeta_\lambda^\alpha, t)$ are coordinates on $\mathscr{V}_\lambda$ where

$$\mathscr{V}_\lambda = \{(\zeta_\lambda, t) \mid |\zeta_\lambda^\alpha| < \varepsilon_\lambda, |t| < \varepsilon\}.$$

Since $\{\mathscr{V}_\lambda\}$ is a refinement of $\{\mathscr{U}_j\}$ we have a map $s: \Lambda \to J$ such that $\mathscr{V}_\lambda \subseteq \mathscr{U}_{s(\lambda)}$. We also have holomorphic transition functions $\varphi_{\lambda\nu}$ where

$$\zeta_\lambda^\alpha = \varphi_{\lambda\nu}(\zeta_\nu, t) \quad \text{on} \quad \mathscr{V}_\lambda \cap \mathscr{V}_\nu.$$

Then the cocycle defined by this covering is

$$\eta_{\lambda\nu}(t) = \sum \frac{\partial \varphi_{\lambda\nu}^\alpha}{\partial t} \frac{\partial}{\partial \zeta_\lambda^\alpha}.$$

As before s induces a map $s^*: \{\theta_{jk}\} \to \{\theta_{\lambda\nu}\}$, where

$$\theta_{\lambda\nu}(t) = r_{\mathscr{V}_\lambda \cap \mathscr{V}_\nu \cap M_t}[\theta_{jk}(t)], j = s(\lambda), k = s(\nu).$$

We must show that $\{\eta_{\lambda\nu}\}$ is cohomologous to $\{\theta_{\lambda\nu}\}$; that is, there exists a cochain $\{\theta_\mu(t)\}$ such that

$$\eta_{\lambda\theta}(t) - \theta_{\lambda\nu}(t) = \theta_\nu(t) - \theta_\lambda(t).$$

Since $\mathscr{V}_\lambda \subseteq \mathscr{U}_j, j = s(\lambda)$, there is a holomorphic g_j such that $z_j^\alpha = g_j^\alpha(\zeta_\lambda, t)$ on $\mathscr{V}_\lambda$. The following equalities are clear:

$$g_j^\alpha[\varphi_{\lambda\nu}(\zeta_\nu, t), t] = g_j^\alpha(\zeta_\lambda, t) = z_j^\alpha$$
$$= f_{jk}^\alpha(z_k, t)$$
$$= f_{jk}^\alpha[g_k(\zeta_\lambda, t), t] \text{ on } \mathscr{V}_\lambda \cap \mathscr{V}_\nu.$$

Differentiating we obtain

$$\sum \frac{\partial g_j^\alpha}{\partial \zeta_\lambda^\beta} \frac{\partial \varphi_{\lambda\nu}^\beta}{\partial t} + \frac{\partial g_j^\alpha}{\partial t} = \sum \frac{\partial f_{jk}^\alpha}{\partial z_k^\beta} \frac{\partial g_k^\beta}{\partial t} + \frac{\partial f_{jk}^\alpha}{\partial t}. \tag{2}$$

Then (2) implies [multiplying by $(\partial/\partial z_j^\alpha)$]

$$\sum \frac{\partial z_j^\alpha}{\partial z_k^\beta}\left(\frac{\partial}{\partial z_j^\alpha}\right) \cdot \frac{\partial g_k^\beta}{\partial t} + \sum \frac{\partial f_{jk}^\alpha}{\partial t} \frac{\partial}{\partial z_j^\alpha} = \sum \frac{\partial z_j^\alpha}{\partial \zeta_\lambda^\beta}\left(\frac{\partial}{\partial z_j^\alpha}\right) \frac{\partial \varphi_{\lambda\nu}^\beta}{\partial t} + \sum \frac{\partial g_j^\alpha}{\partial t} \frac{\partial}{\partial z_j^\alpha}. \tag{3}$$

Hence,

$$\eta_{\lambda\nu} + \sum \frac{\partial g_{s(\lambda)}^\alpha}{\partial t}\left[\frac{\partial}{\partial z_{s(\lambda)}^\alpha}\right] = \sum \frac{\partial g_{s(\nu)}^\beta}{\partial t}\left[\frac{\partial}{\partial z_{s(\nu)}^\beta}\right] + \theta_{\lambda\nu}, \tag{4}$$

on $\mathscr{V}_\lambda \cap \mathscr{V}_\nu$. Therefore if we let

$$\theta_\lambda(t) = \sum \frac{\partial g_{s(\lambda)}^\alpha}{\partial t}\left[\frac{\partial}{\partial z_{s(\lambda)}^\alpha}\right],$$

we get $\eta_{\lambda\nu} - \theta_{\lambda\nu} = \theta_\nu - \theta_\lambda$.   Q.E.D.

So we see that the infinitesimal deformation, $dM_t/dt \in H^1(M_t, \Theta_t)$ is determined uniquely by the family $\mathscr{M} = \{M_t \mid t \in B\}$ and is thus well defined. If we introduce new coordinates on B, $t = t(s)$ so that $t'(s) \neq 0$ then the relation

$$\frac{dM_{t(s)}}{dt} = \frac{dM_t}{dt} \cdot \frac{dt}{ds} \tag{5}$$

is obvious.

Now to return to the more general case, let $\{M_t \mid t \in B\}$ be a family where B is now a general connected complex manifold. Let Δ be a coordinate neighborhood around $b \in B$ and let $(t^1, \cdots, t^m)$ be local coordinates. Then we may assume Δ so chosen that $\bar{\omega}^{-1}(\Delta) = \bigcup_j \mathcal{U}_j$, a union of finitely many coordinate neighborhoods on each of which there are coordinates $(z_j^1, \cdots, z_j^n, t^1, \cdots, t^m)$, where $\mathcal{U}_j = \{(z_j, t) \mid |z_j^\alpha| < \varepsilon_j, t \in \Delta\}$. Again we have transition functions f_{jk}^α

$$z_j^\alpha = f_{jk}^\alpha(z_k, t^1, \cdots, t^m) \text{ on } \mathcal{U}_j \cap \mathcal{U}_k.$$

DEFINITION 3.1. $(\partial M_t/\partial t^\nu) \in H^1(M_t, \Theta_t)$ is the cohomology class of $\{\theta_{jk}\,|_\nu(t)\}$ where

$$\theta_{jk|\nu}(t) = \sum_{\alpha=1}^n \frac{\partial f_{jk}^\alpha(z_k, t)}{\partial t^\nu}\left(\frac{\partial}{\partial z_j^\alpha}\right).$$

If $(\partial/\partial t)$ denotes the tangent vector

$$\frac{\partial}{\partial t} = \sum_{\nu=1}^m c_\nu \frac{\partial}{\partial t^\nu},$$

then we define

$$\frac{\partial M_t}{\partial t} = \sum_{\nu=1}^m c_\nu \frac{\partial M_t}{\partial t^\nu}.$$

We make the following definition:

DEFINITION 3.2. $\mathcal{M} = \{M_t \mid t \in B\}$ is *locally trivial* (complex analytically) if each point $b \in B$ has a neighborhood Δ such that $\bar{\omega}^{-1}(\Delta) = M_b \times \Delta$ (complex analytically). This means that we can choose coordinates (z_j^α, t) such that, $z_j^\alpha = f_{jk}^\alpha(z_k, b)$ (independent of t).

If $\mathcal{M}$ is locally trivial, then each M_t is complex analytically homeomorphic to M_0; hence M_t is independent of t.

PROPOSITION 3.2. If $\mathcal{M}$ is locally trivial then $(\partial M_t/\partial t^\nu) = 0$.

Proof. Trivial.

We mention here a theorem of W. Fischer and H. Grauert (1965).

THEOREM. If each M_t is complex analytically homeomorphic to M_b, then $\mathcal{M}$ is locally trivial.

We now study some examples:

EXAMPLE 1. Let R be a compact Riemann surface. Fix a point $a \in R$. Let w be a coordinate in a neighborhood of a point $b \in R$ such that $w(b) = 0$. We define a family $\{M_t\}$ as follows: M_t will be the branched two-sheeted covering $\tilde{R}_p$ of R with branch points at a and p; $t = w(p)$. We have the question "is $\dfrac{dM_t}{dt} = 0$?" Define the following neighborhoods on R:

$$W_b = \{w \mid |w| < r\},$$
$$W_0 = \{w \mid |w| < r/2\},$$
$$W_1 = \{w \mid r/4 < |w| < r\}.$$

We can write $M_t = U_0 \cup U_1 \cup U_2 \cdots U_j \cdots$ where $U_0 = \pi^{-1}(W_0)$, $U_1 = \pi^{-1}(W_1)$, $\pi(U_j) \cap W_0 = \phi$ for $j \neq 0, 1$ and π is the map $\pi : M_t \to R$ defined by the covering map $\pi : \tilde{R}_p \to R$. We introduce local coordinates as follows on M_t, $(t \in \frac{1}{2}W_0)$:

$$z_0 = \sqrt{w - t} \text{ on } U_0,$$
$$z_1 = \sqrt{w} \quad \text{ on } U_1,$$

and z_j on U_j can be an arbitrary coordinate which should be fixed and independent of t. Then we have $z_j = f_{jk}(z_k, t)$ for holomorphic f_{jk}. In fact,

$$z_0 = f_{01}(z_1, t) = \sqrt{w - t} = \sqrt{z_1^2 - t},$$

and

$$z_j = f_{jk}(z_k) \text{ (independent of } t)$$

for $(j, k) \notin \{(0, 1), (1, 0)\}$. Then $\theta(t) = \{\theta_{jk}(t)\}$ has only one nonzero component,

$$\theta_{01}(t) = \frac{\partial f_{01}}{\partial t} \cdot \left(\frac{\partial}{\partial z_0}\right) = -\frac{1}{2\sqrt{z_1^2 - t}}\left(\frac{\partial}{\partial z_0}\right) = -\frac{1}{2z_0}\left(\frac{\partial}{\partial z_0}\right).$$

Let $V_0 = U_0$, $V_1 = \bigcup_{j \geq 1} U_j$. Then $\theta(t)$ is a 1-cocycle on the covering $\mathscr{V} = \{V_0, V_1\}$; $\theta(t) \in H^1(\mathscr{V}, \Theta_t) \subseteq H^1(M_t, \Theta_t)$.

Suppose $dM_t/dt = 0$. Then there are holomorphic vector fields $\theta_v(t)$ on V_v such that

$$\theta_{01}(t) = \theta_1(t) - \theta_0(t);$$

so

$$\theta_1(t) = -\frac{1}{2z_0}\left(\frac{d}{dz_0}\right) + \theta_0(t). \tag{6}$$

We make the definition

$$\eta(t) = \begin{cases} \theta_1(t) \text{ on } V_1 \\ -\dfrac{1}{2z_0}\left(\dfrac{d}{dz_0}\right) + \theta_0(t) \text{ on } V_0. \end{cases}$$

Then $\eta(t)$ is a vector field on M_t which is holomorphic on $M_t - \{p\}$ and has a simple pole at p.

LEMMA 3.2. If the genus g of R is ≥ 1, then such vector fields η do not exist.

COROLLARY. If $g \geq 1$, then $dM_t/dt \neq 0$, that is, the conformal structure of the branched covering M_t depends on t.

Proof. (of lemma) By the Riemann-Hurwitz formula, we have

$$\chi(M_t) = [2 - 2g(M_t)] = 2\chi(R) - 2$$

where $\chi(M)$ is the Euler characteristic of M. Then the genus $g(M_t)$ equals $2g$. By the Riemann-Roch formula [see Hirzebruch (1962)], there is a holomorphic differential $\varphi(z) = h(z)dz$ on M_t with $2(2g) - 2$ zeros since $2g - 1 \geq 1 > 0$. Since $\eta = y(z)(d/dz)$ has one (simple) pole, $f(z) = h(z)y(z)$ is a meromorphic function on M_t with more zeros than poles $[2(2g) - 2 \geq 2]$. This is impossible (the number of zeros equals the number of poles). Q.E.D.

EXAMPLE 2. Ruled Surfaces (See Chapter 1, Sections 3 and 4.) Recall that $M_t = U_{t_1} \cup U_{t_2}$ where each $U_{t_\nu} = \mathbb{C} \times \mathbb{P}^1$ and

$$(z_1, \zeta_1) \leftrightarrow (z_2, \zeta_2)$$

if and only if

$$\zeta_1 = z_2^m \zeta_2 + t z_2^k, \quad \text{and} \quad z_1 = 1/z_2.$$

We are assuming $m \geq 2k$, $k \geq 1$. Then M_t is independent of $t \neq 0$ for $t \neq 0$ so

$$dM_t/dt = 0 \text{ for } t \neq 0.$$

(For this, one could use the theorem of Fischer and Grauert.) What is $dM_t/dt|_{t=0}$? Consider the covering of M_0, $\mathscr{U} = \{U_{01}, U_{02}\}$; then

$$dM_t/dt|_{t=0} = \theta(0) \in H^1(\mathscr{U}, \Theta_0) \subseteq H^1(M_0, \Theta_0).$$

Then

$$\zeta_1 = f_{12}^1(\zeta_2, z_2, t), \qquad z_1 = f_{12}^2(\zeta_2, z_2),$$

so that

$$\theta_{12}(0) = \left(\frac{\partial f_{12}^1}{\partial t}\right)_{t=0}\left(\frac{\partial}{\partial \zeta_1}\right) = z_2^k\left(\frac{\partial}{\partial \zeta_1}\right) \in \Gamma(U_{01} \cap U_{02}, \Theta).$$

Suppose $dM_t/dt = 0$ at $t = 0$. Then

$$\theta_{12}(0) = \theta_2 - \theta_1$$

where each θ_ν is a holomorphic vector field on $U_{0\nu} = \mathbb{C} \times \mathbb{P}^1$.

LEMMA 3.3. Any holomorphic vector field on $\mathbb{C} \times \mathbb{P}^1$ is of the form

$$\theta = g(z)\left(\frac{\partial}{\partial z}\right) + [a(z)\zeta^2 + b(z)\zeta + c(z)]\left(\frac{\partial}{\partial \zeta}\right),$$

where g, a, b, c are holomorphic functions on $\mathbb{C}$.

Assume Lemma 3.3. We have the following relations:

$$\left(\frac{\partial}{\partial \zeta_2}\right) = z_2^m \left(\frac{\partial}{\partial \zeta_1}\right),$$

$$\left(\frac{\partial}{\partial z_2}\right) = m z_1 \zeta_1 \left(\frac{\partial}{\partial \zeta_1}\right) - z_1^2 \left(\frac{\partial}{\partial z_1}\right),$$

(7)

where (z_ν, ζ_ν) are coordinates on $U_{0\nu} = \mathbb{C} \times \mathbb{P}^1$. Let us compare the co-efficients of $(\partial/\partial \zeta_1)$ in $\theta_2 - \theta_1$ and $\theta_{12}(0)$. From Equations (7), we get

$$z_2^k = z_2^m c_2(z_2) - c_1(z_1)$$

$$= z_2^m c_2(z_2) - c_1\left(\frac{1}{z_1}\right),$$

where the $c_\nu(z_\nu)$ are entire functions. Expanding,

$$z_2^k = \sum_{n=0}^{\infty} c_{2n} z_2^{m+n} - \sum_{n=0}^{\infty} c_{1n} \frac{1}{z_2^n}$$

and $0 < k < m$. This is impossible. Hence,

$$\left.\frac{dM_t}{dt}\right|_{/t=0} \neq 0.$$

For the lemma we have:

Proof. Let $(z, \zeta) \in \mathbb{C} \times \mathbb{P}^1$, where ζ is a nonhomogeneous coordinate on $\mathbb{P}^1$. At $\zeta = \infty.$, the local coordinate on $\mathbb{P}^1$ is $\eta = 1/\zeta$. Restrict the vector field to $\mathbb{C} \times \mathbb{P}^1 - \mathbb{C} \times \{\infty\} = \mathbb{C}^2$. Here

$$\theta = g(z, \zeta)\left(\frac{\partial}{\partial z}\right) + h(z, \zeta)\left(\frac{\partial}{\partial \zeta}\right),$$

where g and h are holomorphic on $\mathbb{C}^2$. At ∞ we have

$$\theta = \gamma(z, \eta)\left(\frac{\partial}{\partial z}\right) + \beta(z, \eta)\left(\frac{\partial}{\partial \eta}\right),$$

where $\zeta = 1/\eta$ and γ, β are holomorphic. Then $\partial\zeta/\partial\eta = -1/\eta^2$ so $(\partial/\partial\eta) = -\zeta^2(\partial/\partial\zeta)$. Hence at ∞,

$$\theta = \gamma(z, \eta)\left(\frac{\partial}{\partial z}\right) - \zeta^2 \beta(z, \eta)\left(\frac{\partial}{\partial \zeta}\right),$$

since $g(z, \zeta) = \gamma(z, \eta), g(z, \xi)$ is holomorphic on $\mathbb{C} \times \mathbb{P}^1$. So $g(z, \zeta)$ is constant as a function of ζ

$$g(z, \zeta) = g(z).$$

Finally, $h(z, \zeta) = -\zeta^2 \beta(z, \eta)$ implies that $h(z, \zeta)$ has a pole of order ≤ 2 at ∞. So $h(z, \zeta) = a(z)\zeta^2 + b(z)\zeta + c(z)$.   Q.E.D.

REMARK 1.   The $\dim_{\mathbb{C}} H^0[M^{(m)}, \Theta]$ is the number of (complex) linearly independent holomorphic vector fields on $M^{(m)}$. We want to compute it. As usual, $M^{(m)} = U_1 \cup U_2$, $U_\nu = \mathbb{C} \times \mathbb{P}^1$, and

$$(z_1, \zeta_1) \leftrightarrow (z_2, \zeta_2)$$

if and only if

$$z_1 = \frac{1}{z_2}, \qquad \zeta_2 = z_2^m \zeta_2.$$

We must count the number of parameters involved in representing a $\theta \in H^0[M^{(m)}, \Theta]$. By the lemma,

$$\theta = \theta_\nu = g_\nu(z_\nu)\left(\frac{\partial}{\partial z^\nu}\right) + a_\nu(z_\nu)\zeta_\nu^2 + b_\nu(z_\nu)\zeta_\nu + c_\nu(z_\nu)\left(\frac{\partial}{\partial \zeta_\nu}\right)$$

on each U_ν, and $\theta_1 = \theta_2$ on $U_1 \cap U_2$. Changing coordinates,

$$\left(\frac{\partial}{\partial \zeta_2}\right) - z_2^m\left(\frac{\partial}{\partial \zeta_1}\right), \qquad \left(\frac{\partial}{\partial z_2}\right) = mz_1\zeta_1\left(\frac{\partial}{\partial \zeta_1}\right) - z_1^2\left(\frac{\partial}{\partial z_1}\right).$$

Hence

$$\theta_2 = -z_1^2 g_2\left(\frac{\partial}{\partial z_1}\right) + mz_1 g_2 \zeta_1\left(\frac{\partial}{\partial \zeta_1}\right)$$

$$+ (a_2 z_1^{2m}\zeta_1^2 + b_2 z_1^m \zeta_1 + c_2)z_2^m\left(\frac{\partial}{\partial z_1}\right)$$

$$= -z_1^2 g_2\left(\frac{\partial}{\partial z_1}\right) + [a_2 z_1^m \zeta_1^2 + (b_2 + mz_1 g_2)\zeta_1 + c_2 z_1^{-m}]\left(\frac{\partial}{\partial \zeta_1}\right)$$

$$= \theta_1 = g_1\left(\frac{\partial}{\partial z_1}\right) + (a_1\zeta_1^2 + b_1\zeta_1 + c_1)\left(\frac{\partial}{\partial \zeta_1}\right).$$

Equating coefficients,

$$g_1(z_1) = -z_1^2 g_2(z_2), \; a_1(z_1) = z_1^m a_2(z_2),$$

$$b_1(z_1) = b_2(z_2) + mz_1 g_2(z_2), \; c_1(z_1) = z_1^{-m} c_2(z_2).$$

These functions are all entire functions of z_1. Let us investigate their behavior at $z_1 = \infty$. Since $z_2 = 1/z_1$, we see that g_1 has a pole of order ≤ 2 at ∞, a_1 has a pole of order $\leq m$ at ∞ and c_1 has a zero at ∞. Assume that $m \geq 1$. Then

$$g_1 = g_{10} z_1^2 + g_{11} z_1 + g_{12},$$
$$a_1 = a_{10} z_1^m + \cdots + a_{1m},$$
$$c_1 = 0 \quad \text{(by Liouville's theorem)}.$$

Consider the b terms:

$$b_1 = \sum_{n=0}^{\infty} b_{1n} z_1^n = \sum_{n=0}^{\infty} b_{2n} \frac{1}{z_1^n} - mg_{10} z_1 - mg_{11} - mg_{12} \frac{1}{z_1}.$$

So $b_1(z_1) = -mg_{10} z_1 b_{10}$. θ depends linearly on $(g_{10}, g_{11}, g_{12}, a_{10}, \cdots, a_{1m}, b_{10})$. Hence,

$$\dim_{\mathbb{C}} H^0[M^{(m)}, \Theta] = m + 5. \tag{8}$$

We therefore have:

THEOREM 3.1. $M^{(m)} \neq M^{(n)}$ (complex analytically) if $n \neq m$.

REMARK 2. $M^{(2n)} \neq M^{(2n-1)}$ *topologically.*

REMARK 3. Let $\{M_t \mid t \in \mathbb{C}\}$, M_t given by

$$\zeta_1 = z_2^m \zeta_2 + t z_2^k, \quad z_1 = \frac{1}{z_2}$$

as before. Then $M_0 = M^{(m)}$, $M_t = M^{(m-2k)}$ for $t \neq 0$. And we have shown

$$dM_t/dt = \begin{cases} 0, & \text{for } t \neq 0 \\ \neq 0, & \text{for } t = 0. \end{cases}$$

Suppose we "reparametrize" and consider $\{M_{s^2} \mid s \in \mathbb{C}\}$. M_{s^2} is defined by

$$\zeta_1 = z_2 \zeta_2 + s^2 z_2^k, \qquad z_1 = \frac{1}{z_2}.$$

Then $M_0 = M^{(m)}$, $M_{s^2} = M^{(m-2k)}$, $s \neq 0$ as above. *But*

$$\frac{dM_{s^2}}{ds} = \frac{dM_t}{dt} \cdot \frac{ds^2}{ds} = \frac{2s\, dM_t}{dt} = 0$$

for all $s \in \mathbb{C}$. We know that M_t independent of t implies $dM_t/dt = 0$. We have just seen that $dM_t/dt = 0$ does not imply that M_t is independent of t.

However, we have the following theorems:

THEOREM α. If dim $H^1(M_t, \Theta_t)$ is independent of t and if $\partial M_t/\partial t^\nu = 0$ for all ν and t, then $\{M_t \mid t \in B\}$ is locally trivial and hence M_t is independent of t.

THEOREM β. The function $t \to H^1(M_t, \Theta_t)$ is an upper semicontinuous function of t. That is

$$\dim H^1(M_t, \Theta_t) \leq H^1(M_s, \Theta_s),$$

if t is in a sufficiently small neighborhood of s; that is,

$$\varlimsup_{t \to s} \dim H^1(M_t, \Theta_t) \leq H^1(M_s, \Theta_s).$$

THEOREM γ. If $H^1(M_s, \Theta_s) = 0$, then $M_t = M_s$ for t in a small neighborhood of s.

Theorem α is proved in Kodaira and Spencer (1958a), Theorem β in Kodaira and Spencer (1960), and Theorem γ is due to Frölicher and Nijenhuis (1951). Theorem β follows from some results which we will prove in a later chapter. Theorem α will not be proved here.

DEFINITION 3.3. We say that a compact complex manifold M is *rigid* if, for any complex analytic family $\{M_t \mid t \in B\}$ such that $M_{t_0} = M$, we can find a neighborhood N of t_0 such that $M_t = M_{t_0}$ for $t \in N$. (More precisely, if $\varpi : \mathcal{M} \to B$ is the family $\{M_t\}$, then $\varpi^{-1}(N) = N \times M_{t_0}$ complex analytically.)

The following theorem follows from Theorem γ.

THEOREM 3.2. If $H^1(M, \Theta) = 0$, then M is rigid. We will give a proof of this using elementary methods. We have the following:

PROBLEM. Find an example of an M which is rigid, but $H^1(M, \Theta) \neq 0$. (Not easy?)

REMARK. $\mathbb{P}^n$ is rigid. For $n \geq 2$ the only known proof is to show $H^1(\mathbb{P}^n, \Theta) = 0$ [Bott (1957)]. Let us proceed to the proof.

Proof. (of Theorem 3.2) The proof will be elementary in that it consists of two elementary ideas:

(1) Construction of a formal power series, and
(2) proof of convergence.

The proof is actually long and computational, so please stay with us. It makes no difference for the proof and it makes the writing much easier if we assume dim $B = 1$. The result is local so we may assume $B = \{t \mid |t| < r\}$ and $t_0 = 0$. We can cover $\bar{\omega}^{-1}(\Delta_\varepsilon)$, $\Delta_\varepsilon = \{t \mid |t| < \varepsilon\}$ with coordinates

$$\mathcal{U}_j = \{(z_j, t) \mid |z_j^\alpha| < \varepsilon_j, |t| < \varepsilon\}.$$

Then

$$z_j^\alpha = f_{jk}^\alpha(z_k, t) \quad \text{on} \quad \mathcal{U}_j \cap \mathcal{U}_k.$$

M is covered by $\cup_j U_j^0 = M$ where

$$U_j^0 = \{z_j \mid |z_j^\alpha| < \varepsilon_j\} \times \{0\} \subseteq \mathcal{U}_j.$$

Then $M \times B = \cup_j (U_j^0 \times B)$ where for $(w_\nu, t) \in U_\nu^0 \times B$,

$$(w_j, t) \leftrightarrow (w_k, t)$$

if and only if

$$w_j^\alpha = f_{jk}^\alpha(w_k, 0);$$

that is,

$$w_j^\alpha = g_{jk}^\alpha(w_k), \qquad \text{where } g_{jk}^\alpha(w_k) = f_{jk}^\alpha(w_k, 0). \tag{9}$$

We can rephrase our result:

THEOREM. If δ is sufficiently small there is a biholomorphic map φ of $\bar{\omega}^{-1}(\Delta_\delta)$ onto $M \times \Delta_\delta$ such that φ: maps $\bar{\omega}^{-1}(t)$ onto $M \times t$ and $\varphi: M = \bar{\omega}^{-1}(0) \to M \times 0$ is the identity map.

Suppose we choose δ so that φ maps

$$\mathcal{U}_k^\delta = \{(z_k, t) \mid |z_k^\alpha| < \varepsilon_k', |t| < \delta\} \qquad 0 < \varepsilon_k' < \varepsilon_k$$

into $U_k^0 \times B$, $\varphi(\mathcal{U}_k^\delta) \subseteq U_k^0 \times B$. Let $(z_j, t) \in \mathcal{U}_j^\delta$. Then, $\varphi(z_j, t) = (w_j, t) = [\varphi_j(z_j, t), t]$ so on each $\mathcal{U}_j^\delta$, φ is represented by holomorphic functions $\varphi_j^\delta(z_j, t)$ where $\varphi_j^\alpha(z_j, 0) = z_j^\alpha$. On $\mathcal{U}_j^\delta \cap \mathcal{U}_k^\delta$,

$$z_j^\alpha = f_{jk}^\alpha(z_k, t);$$

so

$$\varphi_j^\alpha(z_j, t) = g_{jk}^\alpha[\varphi_k(z_k, t)]$$

implies

$$\varphi_j^\alpha[f_{jk}(z_k, t), t] = g_{jk}^\alpha[\varphi_k(z_k, t)]. \tag{10}$$

Therefore we see that we can prove the theorem if we can construct holomorphic functions $\varphi_j^\alpha(z_j, t)$ on $\mathcal{U}_j^\delta$ satisfying (10) and

$$\varphi_j^\alpha(z_j, 0) = z_j^\alpha. \tag{11}$$

For simplicity we may as well assume that $\mathcal{U}_j$ is of the form

$$\mathcal{U}_j = \{(z_j, t)\mid |z_j^\alpha| < 1, |t| < \varepsilon\}, \quad M = \bigcup U_j^0,$$

$U_j^0 = \{z_j\mid |z_j^\alpha| < 1 + v \text{ for some } v > 0\}$ and $U_j^0 \supset M \cap \mathcal{U}_j$. If expand $\varphi_j(z_j, t)$ into a power series, we get

$$\varphi_j(z_j, t) = z_j + \varphi_{j|1}(z_j)t + \varphi_{j|2}(z_j)t^2 + \cdots + \varphi_{j|m}(z_j)t^m + \cdots, \tag{12}$$

where each $\varphi_{j|m}(z_j)$ is a holomorphic vector valued function. If we expand both sides of (10) we get

$$\sum_{m=0}^{\infty} F_m(\varphi_{j|1}, \cdots, \varphi_{j|m}) = \sum_{m=0}^{\infty} G_m(\varphi_{k|1}, \cdots, \varphi_{k|m})t^m, \tag{13}$$

where F_m and G_m are polynomials. We introduce some notation: If $P(t) = \sum P_n t^n$ and $Q(t) = \sum Q_n t^n$ are two power series, $P(t) \underset{m}{\equiv} Q(t)$ means $Q_n = P_n$ up to $n = m$ [that is, $P(t) \equiv Q(t) \bmod (t^{m+1})$]. Therefore, to solve (formally) Equation (13) we need only solve

$$\varphi_j^m[f_{jk}(z_k, t), t] \underset{m}{\equiv} g_{jk}[\varphi_k^m(z_k, t)], \tag{$13)_m$}$$

for each m, where

$$\varphi_j^m(z_j, t) = z_j + \cdots + \varphi_{j|m}(z_j)t^m. \tag{14}$$

First consider $m = 1$. We have

$$z_j = f_{jk}(z_k, t) = g_{jk}(z_k) + \sum_{m=1}^{\infty} f_{jk|m}(z_k)t^m.$$

Using $(13)_1$,

$$g_{jk}(z_k) + f_{jk|1}(z_k)t + \varphi_{j|1}[g_{jk}(z_k)]t \underset{1}{\equiv} g_{jk}[z_k + \varphi_{k|1}(z_k)t]$$

$$\underset{1}{\equiv} g_{jk}(z_k) + \sum_{\beta=1}^{n} \frac{\partial g_{jk}(z_k)}{\partial z_k^\beta} \varphi_{k|1}^\beta(z_k)t.$$

So

$$f_{jk|1}^\alpha(z_k) = \sum_{\beta=1}^{n} \frac{\partial z_j^\alpha}{\partial z_k^\beta} \varphi_{k|1}^\beta(z_k) - \varphi_{j|1}^\alpha(z_j). \tag{$13)_1$}$$

Now

$$\theta_{jk} = \sum_\alpha \left(\frac{\partial f_{jk}^\alpha}{\partial t}\right)_{t=0} \left(\frac{\partial}{\partial z_j^\alpha}\right) = \sum f_{jk|1}^\alpha \left(\frac{\partial}{\partial z_j^\alpha}\right)$$

belongs to $H^1(M, \Theta)$. By assumption $H^1(M, \Theta) = 0$ so $\{\theta_{jk}\}$ is cohomologous to zero. But Equation $(13)_1$ says we must find $\{\varphi_{k|1}\}$ so that $\theta_{jk} = \varphi_{k|1} - \varphi_{j|1}$. $H^1(M, \Theta) = 0$ allows us to do this so the first step in an induction proof is completed.

Assume that $\varphi_j^m(z_j, t)$ are determined so that Equation $(13)_m$ holds, that is, $\varphi_j^m(f_{jk}) - g_{jk}(\varphi_k^m) \underset{m+1}{\equiv} \Gamma_{jk} t^{m+1}$. We must show that $\varphi_{j\,|\,m+1}(z_j)$ can be determined so that $\varphi_j^{m+1} = \varphi_j^m + \varphi_{j\,|\,m+1} t^{m+1}$ satisfies $(13)_{m+1}$, which is

$$\varphi_j^{m+1}[f_{jk}(z_k, t), t] - g_{jk}(\varphi_k^{m+1}) \underset{m+1}{\equiv} 0.$$

This is equivalent to

$$\varphi_j^m(f_{jk}) + \varphi_{j|m+1}[f_{jk}(z_k, t)]t^{m+1} \underset{m+1}{\equiv} g_{jk}[\varphi_k^m(z_k, t) + \varphi_{k|m+1}(z_k) \cdot t^{m+1}].$$

We use

$$\varphi_{j|m+1}[f_{jk}(z_k, t)]t^{m+1} \underset{m+1}{\equiv} \varphi_{j|m+1}[f_{jk}(z_k, 0)]t^{m+1}$$

and $f_{jk}(z_k, 0) = g_{jk}(z_k) = z_j$ to get

$$\varphi_j^m(f_{jk}) + \varphi_{j|m+1}(z_j)t^{m+1} \underset{m+1}{\equiv} g_{jk}^\alpha(\varphi_k^m) + \sum \frac{\partial z_j^\alpha}{\partial z_k^\beta} \varphi_{k|m+1}^\beta(z_k)t^{m+1}.$$

Here we have used

$$g_{jk}^\alpha(\varphi_k^m + y) = g_{jk}^\alpha(\varphi_k^m) + \sum_\beta \frac{\partial g_{jk}^\alpha}{\partial y^\beta}(\varphi_k^m)y^\beta + \sum (\text{---})y^\beta y^\nu + \cdots,$$

$$g_{jk}^\alpha(\varphi_k^m + \varphi_{k|m+1}t^{m+1}) \underset{m+1}{\equiv} g_{jk}^\alpha(\varphi_k^m) + \sum \frac{\partial g_{jk}^\alpha}{\partial y^\beta}[\varphi_k^m(z_k, t)]\varphi_{k|m+1}^\beta t^{m+1},$$

and $\varphi_k^m(z_k, t) = z_k + \cdots$. So if we can solve

$$\Gamma_{jk}^\alpha(z_j) = \sum_\beta \frac{\partial z_j^\alpha}{\partial z_k^\beta} \varphi_{k|m+1}(z_k) - \varphi_{j|m+1}^\alpha(z_j), \tag{15}$$

for $\varphi_{j\,|\,m+1}^\nu$ we will have $(13)_{m+1}$. Let

$$\Gamma_{jk} = \sum_\alpha \Gamma_{jk}^\alpha(z_j)\left(\frac{\partial}{\partial z_j^\alpha}\right),$$

$$\varphi_{j|m+1} = \sum_\alpha \varphi_{j|m+1}^\alpha(z_j)\left(\frac{\partial}{\partial z_j^\alpha}\right).$$

Then we want to solve $\Gamma_{jk} = \varphi_{k\,|\,m+1} - \varphi_{j\,|\,m+1}$. We claim that $\{\Gamma_{jk}\}$ is a cocycle [belongs to $H^1(M, \Theta)$]. Then we would be done as before, since $H^1(M, \Theta) = 0$ and $\{\Gamma_{jk}\}$ must be a coboundary, $\Gamma_{jk} = \theta_k - \theta_j$, and set $\theta_{j\,|\,m+1} = \theta_j$.

LEMMA 3.4. $\{\Gamma_{jk}\}$ is a cocycle, that is, $\Gamma_{ik} = \Gamma_{ij} + \Gamma_{jk}$ on $\mathcal{U}_i \cap \mathcal{U}_j \cap \mathcal{U}_k \cap M$. That is,

$$\Gamma_{ik}^\alpha(z_i) = \Gamma_{ij}^\alpha(z_i) + \sum_{\beta=1}^n \frac{\partial z_i^\alpha}{\partial z_j^\beta} \Gamma_{jk}^\beta(z_j).$$

Proof. By definition

$$\Gamma_{ik}(z_i)t^{m+1} \underset{m+1}{\equiv} \varphi_i^m[f_{ik}(z_k, t), t] - g_{ik}[\varphi_k^m(z_k, t)],$$

so

$$g_{ik}(\varphi_k^m) = g_{ij}[g_{jk}(\varphi_k^m)] \underset{m+1}{\equiv} g_{ij}[\varphi_j^m(f_{jk}) - \Gamma_{jk}t^{m+1}].$$

Now $\Gamma_{jk}t^{m+1} \underset{m+1}{\equiv} \varphi_j^m(f_{jk}) - g_{jk}(\varphi_k^m)$ and

$$g_{ij}[\varphi_j^m(f_{jk}) - \Gamma_{jk}t^{m+1}] \underset{m+1}{\equiv} g_{ij}^\alpha[\varphi_j^m(f_{jk})] - \sum_\beta \frac{\partial g_{ij}^\alpha}{\partial z_j^\beta} \{\varphi_j^m[f_{jk}(z_k, t)]\}\Gamma_{jk}^\beta t^{m+1}.$$

So

$$\Gamma_{ik}(z_i)t^{m+1} \underset{m+1}{\equiv} \varphi_i^m[f_{ik}(z_k, t), t] - g_{ij}[\varphi_j^m(f_{jk}, t)] - \sum_\beta \frac{\partial z_i^\alpha}{\partial z_j^\beta} \Gamma_{jk}^\beta t^{m+1}$$

$$\underset{m+1}{\equiv} \varphi_i^m\{f_{ij}[f_{jk}(z_k, t), t], t\} - g_{ij}[\varphi_j^m(f_{jk}, t)] - \sum \frac{\partial z_i^\alpha}{\partial z_j^\beta} \Gamma_{jk}^\beta t^{m+1}.$$

By assumption

$$\varphi_i^m[f_{ij}(z_j, t), t] - g_{ij}[\varphi_j^m(z_j, t)] \underset{m+1}{\equiv} \Gamma_{ij}(z_j)t^{m+1},$$

$$\underset{m+1}{=} \Gamma_{ij}[f_{jk}(z_k, t)]t^{m+1} - \cdots.$$

Hence

$$\Gamma_{ik}^\alpha(z_i) = \Gamma_{ij}^\alpha(z_j) + \sum \frac{\partial z_i^\alpha}{\partial z_j^\beta} \Gamma_{jk}^\beta(z_j). \qquad \text{Q.E.D.}$$

This finishes the construction of the formal power series

$$\varphi_j(z_j, t) = z_j + \varphi_{j|1}(z_j)t + \cdots,$$

$$z_j \in \mathcal{U}_j \cap M, \quad \mathcal{U}_j \cap M = \{z_j|\ |z_j|^\alpha < 1\}$$

such that $\varphi_j[f_{jk}(z_k, t)] = g_{jk}[\varphi_k(z_k, t)]$ as formal power series.

LEMMA 3.5. The power series $\varphi_j(z_j, t)$ converges for $|t| < \delta$ for some small $\delta > 0$.

Proof. We dominate φ_j with a convergent series. We fix some notation. Let $\psi(z, t) = \sum_{m=0}^\infty \psi_m(z)t^m$ be a power series where $\psi_m(z) = [\psi_m^1(z), \cdots, \psi_m^n(z)]$ $z \in U$. Let $a(t) = \sum_{m=0}^\infty a_m t^m$, $a_m \geq 0$ be a series with real, positive coefficients. We write $\psi(z, t) \ll a(t)$ and say that $a(t)$ *dominates* $\psi(z, t)$ if

$|\psi_m^\alpha(z)| \leq a_m$ for all $z \in U$ and all $\alpha = 1, \cdots, n$. The *norm* of ψ_m is $|\psi_m| = \max\limits_{\alpha} \sup\limits_{z \in U} |\psi_m^\alpha(z)|$. Consider the series

$$A(t) = \frac{b}{16c} \sum_{m=1}^{\infty} \frac{(ct)^m}{m^2},$$

where b and c are constants to be determined later. Then

$$A(t) = \frac{b}{16} \left\{ t + \sum_{m=2}^{\infty} \frac{c^{m-1} t^m}{m^2} \right\}$$

converges for $|t| < 1/c$. In Lemma 3.5 it suffices to prove

$$\varphi_j(z_j, t) - z_j \ll A(t). \tag{16}$$

In fact, it suffices to prove

$$\varphi_j^m(z_j, t) - z_j \ll A(t) \qquad \text{for } m = 1, 2, 3, \cdots. \tag{$16)_m$}$$

First, we prove:

LEMMA 3.6. $[A(t)]^2 \ll (b/c)A(t)$.

Proof. $[A(t)]^2 = \left(\dfrac{b}{16c}\right)^2 \left[\sum\limits_{m=1}^{\infty} \dfrac{(ct)^m}{m^2} \right] \left[\sum\limits_{k=1}^{\infty} \dfrac{(ct)^k}{k^2} \right]$

$\qquad\qquad\quad = \left(\dfrac{b}{16c}\right)^2 \sum\limits_{n=2}^{\infty} (ct)^n \sum\limits_{m+k=n} \dfrac{1}{m^2 k^2}.$

Since

$$\sum_{m+k=n} \frac{1}{m^2 k^2} \leq 2 \sum_{\substack{k+m=n \\ m \leq k}} \frac{1}{m^2 k^2},$$

and $m \leq k$ implies $k \geq n/2$, we get

$$\sum_{m+k=n} \frac{1}{m^2 k^2} \leq \frac{2}{(n/2)^2} \sum_{m=1}^{\infty} \frac{1}{m^2} \leq \frac{8}{n^2} \frac{\pi^2}{6} < \frac{16}{n^2}.$$

Hence,

$$[A(t)]^2 \ll \left(\frac{b}{16c}\right)^2 \sum_{n=2}^{\infty} \frac{(ct)^n}{n^2} \, 16 \ll \frac{b}{c} \frac{b}{16c} \sum_{n=1}^{\infty} \frac{(ct)^n}{n^2}. \qquad \text{Q.E.D.}$$

COROLLARY. $A^n(t) \ll (b/c)^{n-1} A(t)$.

Let us prove $(16)_1$: We want to show that

$$\varphi_{j|1}(z_j)t \ll A(t) = \frac{b}{16} \{ t + \cdots \}.$$

It is enough to prove $|\varphi^\alpha_{j\,|\,1}(z_j)| \leq b/16$. From $(13)_1$,

$$\sum \frac{\partial z^\alpha_j}{\partial z^\beta_k} \varphi^\beta_{k|1}(z_k) - \varphi^\alpha_{j|1}(z_j) = f^\alpha_{jk|1}(z_k).$$

We may (perhaps by shrinking U_m to U^*_m so that $\cup\, U^*_m = M$) assume that the given functions $f_{jk\,|\,1}$ are bounded. Also the $\varphi_{j\,|\,1}(z_j)$ are holomorphic on U^*_k and we may, therefore, assume a finite upper bound for all of them. (Compactness is needed for these statements.) So we see $(16)_1$ is satisfied for b large enough.

By induction, assume $(16)_m$ and let us prove $(16)_{m+1}$. Remember

$$\varphi^{m+1}_j = \varphi^m_j + \varphi_{j\,|\,m+1}(z_j)t^{m+1},$$

$$\Gamma^\alpha_{jk}(z_j) = \sum_{\beta=1}^n \frac{\partial z^\alpha_j}{\partial z^\beta_k} \varphi^\beta_{k|m+1}(z_k) - \varphi^\alpha_{j|m+1}(z_j),$$

and

$$\Gamma_{jk}(z_k)t^{m+1} \underset{m+1}{\equiv} \varphi^m_j[f_{jk}(z_k,t),t] - g_{jk}[\varphi^m_k(z_k,t)].$$

We fix some more notation: For $\psi(t) = \sum \psi_m t^m$, let

$$[\psi(t)]_{m+1} = \psi_{m+1} t^{m+1}.$$

Then

$$\Gamma_{jk} t^{m+1} = [\varphi^m_j[f_{jk}(z_k,t)]]_{m+1} - [g_{jk}[\varphi^m_k(z_k,t)]]_{m+1}.$$

Remember the definitions of U_j and U^0_j, $U_j \subset U^0_j$. Then $g_{jk}(z_k)$ is defined on $U^0_j \cap U^0_k$ and $g_{jk}(z_k + y)$, $y = (y_1, \cdots, y_n)$ is holomorphic for $z_k \in U_j \cap U_k$, $|y| < v$. So there is a $\kappa > 0$ so that

$$g_{jk}(z_k + y) \ll \sum_{m=0}^\infty \kappa^m(y_1 + \cdots + y_n)^m. \tag{17}$$

For the moment, let $\psi_k(z_k,t) = \varphi^m_k(z_k,t) - z_k$. We want to estimate $[g_{jk}(\varphi^m_k)]_{m+1} = [g_{jk}(z_k + \psi_k)]_{+m1}$ where $m + 1 \geq 2$. But,

$$[g_{jk}(z_k + \psi_k)]_{m+1} = \left[g_{jk}(z_k + \psi_k) - z_j - \sum \frac{\partial g_{jk}}{\partial z^\beta_k}(z_k)\psi^\beta_k(z_k,t)\right]_{m+1}$$

since $\psi_k(z_k,t)$ is a polynomial of degree $\leq m$ in t. From Equation (17), we get

$$g_{jk}(z_k + y) - z_j - \sum \frac{\partial g_{jk}}{\partial z^\beta_k} y^\beta \ll \sum_{r=2}^\infty \kappa^r(y_1 + \cdots + y_n)^r,$$

hence,

$$[g_{jk}(z_k + \psi_k)]_{m+1} \ll \sum_{r=2}^{\infty} \kappa^r [nA(t)]^r$$

$$\ll A(t) \sum_{r=2}^{\infty} \kappa^r n^r \left(\frac{b}{c}\right)^{r-1}$$

$$= A(t) \frac{\kappa^2 n^2 b}{c} \frac{1}{1 - (\kappa nb/c)},$$

because by induction $\psi_k \ll A(t)$. If we choose c so large that $(\kappa nb/c) < \frac{1}{2}$, then

$$\{g_{jk}[\varphi_k^m(z_k, t)]\}_{m+1} \ll \frac{2\kappa^2 n^2 b}{c} A(t). \tag{18}$$

Now we want to estimate

$$\Gamma_{jk}(z_j)t^{m+1} \underset{m+1}{\equiv} \varphi_j^m[f_{jk}(z_k, t), t] - g_{jk}[\varphi_k^m(z_k, t)].$$

Remember that Γ is a cocycle so

$$\Gamma_{ik}^{\alpha} = \Gamma_{ij}^{\alpha} + \sum \frac{\partial z_i^{\beta}}{\partial z_j^{\beta}} \Gamma_{jk}^{\alpha} \text{ on } U_i \cap U_j \cap U_k,$$

where U_i satisfies $U_i = \{z_i | \, |z_i| < 1\}$. We choose $U_i^* \subset U_i$, $U_i^* = \{z_i | \, |z_i| < 1 - \beta\}$ such that $\cup U_i^* = M$. It suffices to estimate $\Gamma_{ik}(z_i)$ for $z \in U_i^* \cap U_i$, for take any $z \in U_j \cap U_k$, then $z \in U_i^*$ for some i, and

$$\Gamma_{jk}^{\beta}(z_j) = \sum_{\alpha} \frac{\partial z_j^{\beta}}{\partial z_i^{\alpha}} \{\Gamma_{ik}^{\alpha}(z_i) - \Gamma_{ij}^{\alpha}(z_i)\}.$$

Then if $\Gamma_{jk}(z_j)t^{m+1} \ll B(t)$ for $z \in U_j^* \cap U_k$, $\Gamma_{jk}(z_j)t^{m+1} \ll K_1 B(t)$ for $z \in U_j^* \cap U_k$, where

$$K_1 = 2n \max_{i, j} \left(\sup \left| \frac{\partial z_j^{\alpha}}{\partial z_i^{\beta}} \right| \right).$$

So consider $\varphi_j^m[f_{jk}(z_k, t), t]$ for $z \in U_j^* \cap U_k$. Expanding f_{jk},

$$f_{jk}(z_k, t) = z_j + f_{jk|1}(z_k)t + \cdots,$$

we see we may assume that

$$\sigma_{jk}(z_k, t) = f_{jk}(z_k, t) - z_j \ll A_0(t) = \frac{b_0}{16c_0} \sum_{m=1}^{\infty} \frac{(c_0 t)^m}{m^2},$$

for small t and some constants b_0, c_0. Our induction hypothesis is

$$\varphi_j^m - z_j = \sum_{\mu=1}^{m} \varphi_{j|\mu}(z_j)t^{\mu} \ll A(t) = \sum_{1}^{\infty} a_{\mu} t^{\mu}.$$

Since each $\varphi_{j\mid\mu}$ is holomorphic, we get a power series

$$\varphi_{j\mid\mu}(z_j + y) - \varphi_{j\mid\mu}(z_j) = \sum_{\nu_1 + \cdots + \nu_n \geq 1} c_{\nu_1 \cdots \nu_n} y_1^{\nu_1} \cdots y_n^{\nu_n},$$

where

$$c_{\nu_1 \cdots \nu_n} = \left(\frac{1}{2\pi_i}\right)^n \int \cdots \int_{|y_\alpha| = \beta} \frac{\varphi_{j\mid\mu}(z_j + y)}{y_1^{\nu_1 + 1} \cdots y_n^{\nu_n + 1}} \, dy_1 \cdots dy_n.$$

Since $|\varphi_{j\mid\mu}| < a_\mu$ for some constant a_μ, on $|y_\alpha| = \beta$, we see that

$$|c_{\nu_1 \cdots \nu_n}| \leq (a_\mu / \beta^{\nu_1 \cdots \nu_n}).$$

We thus obtain

$$\varphi_{j\mid\mu}(z_j + y) - \varphi_{j\mid\mu}(z_j) \ll a_\mu \sum_{\nu_1 + \cdots + \nu_n \geq 1} \frac{y_1^{\nu_1} \cdots y_n^{\nu_n}}{\beta^{\nu_1 + \cdots + \nu_n}}.$$

Summing we get

$$\varphi_j^m(z_j + y, t) - \varphi_j^m(z_j, t) \ll \sum_{\mu \geq 1} a_\mu t^\mu \cdot \sum_{\nu_1 + \cdots + \nu_n \geq 1}^{\infty} \frac{y_1^{\nu_1} \cdots y_n^{\nu_n}}{\beta^{\nu_1 + \cdots + \nu_n}}. \tag{19}$$

Again, since $\varphi_j^m(z_j, t)$ only has terms of degree $\leq m$ in t^m we obtain

$$[\varphi_j^m[f_{jk}(z_k, t), t]]_{m+1} = [\varphi_j^m(z_j, \sigma_{jk}, t) - \varphi_j^m(z_j, t)]_{m+1}.$$

Using (19) we get

$$[\varphi_j^m[f_{jk}(z_k, t), t]]_{m+1} \ll A(t) \sum_{\nu_1 + \cdots + \nu_n \geq 1} \left[\frac{A_0(t)}{\beta}\right]^{\nu_1 + \cdots + \nu_n}$$

$$= A(t)\left\{\left[\sum_{\nu=0}^{\infty} \left(\frac{A_0}{\beta}\right)^\nu\right]^n - 1\right\}.$$

The corollary to Lemma 3.6 gives $A_0(t)^\nu \leq (b_0/c_0)^{\nu-1} A_0(t)$, so

$$\left[\sum_{\nu=0}^{\infty} \left(\frac{A_0}{\beta}\right)^\nu\right]^n - 1 \leq \left[1 + \sum_{\nu=1}^{\infty} \frac{1}{\beta^\nu}\left(\frac{b_0}{c_0}\right)^{\nu-1} A_0(t)\right]^n - 1$$

$$= \left[1 + \frac{1}{\beta}\frac{1}{1 - (b_0/\beta c_0)} A_0(t)\right]^n - 1.$$

Since we have chosen c_0 so that $A_0(t) \gg f_{jk}(z_k, t) - z_k$, we may replace c_0 with a larger constant and assume that $b_0/2c_0 < \frac{1}{2}$. Then

$$\left[\sum_{\nu=0}^{\infty} \left(\frac{A_0}{\beta}\right)^\nu\right]^n - 1 \ll \left[1 + \frac{2}{\beta} A_0(t)\right]^n - 1$$

$$= \frac{2n}{\beta} A_0 + \binom{n}{2}\left(\frac{2A_0}{\beta}\right)^2 + \cdots + \left(\frac{2A_0}{\beta}\right)^n$$

$$\ll \frac{K_2}{\beta} A_0(t),$$

where K_2 is a constant depending only on n, β, b_0, and c_0. Thus,

$$[\varphi_j^m[f_{jk}(z_k, t), t]]_{m+1} \ll \frac{K_2}{\beta} A(t)A_0(t).$$

We may assume $b > b_0$, $c > c_0$ and then $A(t) \gg A_0(t)$, so

$$\frac{K_2}{\beta} A_0(t)A(t) \ll \frac{K_2}{\beta} \frac{b}{c} A(t).$$

Finally,

$$\Gamma_{jk} t^{m+1} = [\varphi_j^m[f_{jk}(z_k, t), t]]_{m+1} - [g_{jk}[\varphi_k(z_k, t)]]_{m+1}$$

$$\ll \left(\frac{K_2}{\beta} - 2\kappa^2 n^2\right) \frac{b}{c} A(t),$$

for $z \in U_j^* \cap U_k$. Hence

$$\Gamma_{jk} t^{m+1} \ll K_1 \left(\frac{K_2}{\beta} + 2\kappa^2 n^2\right) \frac{b}{c} A(t), \tag{20}$$

for $z \in U_j \cap U_k$. Now

$$\Gamma_{jk}^\alpha = \sum \frac{\partial z_j^\alpha}{\partial z_k^\beta} \varphi_{k|m+1}^\beta(z_k) - \varphi_{j|m+1}^\alpha(z_j)$$

and we want to estimate $\varphi_{j|m+1}^\alpha(z_j)$. As before, consider $\Gamma_{jk} = \sum \Gamma_{jk}^\alpha(z_j)(\partial/\partial z_j^\alpha)$, $\varphi_{j|m+1} = \sum \varphi_{j|m+1}^\alpha(\partial/\partial z_j^\alpha)$. With these notations, $\Gamma_{jk} = \varphi_{k|m+1} - \varphi_{j|m+1}$. At this point we need a lemma which plays a crucial role in arguments of this type. Let $\Gamma = \{\Gamma_{jk}\}$ be a 1-cocycle where Γ_{jk} is a holomorphic vector field on $U_j \cap U_k$. Let $\psi = \{\psi_j\}$ be an 0-cochain where ψ_j is a holomorphic vector field on U_j. We define

$$\|\Gamma\| = \max_{j,k} \sup_{z \in U_j \cap U_k} \max_\alpha |\Gamma_{jk}^\alpha(z_j)|.$$

$$\|\psi\| = \max_j \sup_{z \in U_j} \max_\alpha |\psi_j^\alpha(z_j)|.$$

LEMMA 3.7. There exists a constant K such that if Γ is cohomologous to 0, then we can find ψ with $\delta\psi = \Gamma$ satisfying $\|\psi\| \leq K\|\Gamma\|$ where M is a compact complex manifold, $H^1(M, \Theta)$ need not be zero, and K does not depend on Γ.

Proof. Remember $U_j = \{z_j| |z_j^\alpha| < 1\}$ and $U_j^*\{z_j| |z_j^\alpha| < 1 - \beta\}$. Suppose the lemma is not true. Let

$$\tau(\Gamma) = \{\|\psi\| \,|\, \delta\psi = \Gamma\} < +\infty.$$

If $\tau(\Gamma) \leq K\|\Gamma\|$, then there is ψ such that $\delta\psi = \Gamma$ and $\|\psi\| \leq (K+1)\|\Gamma\|$. Thus if the lemma is false, $\tau(\Gamma)/\|\Gamma\|$ is unbounded. Hence there is a sequence $\{\Gamma^{(\nu)}\}$

such that $\tau[\Gamma^{(\nu)}]/\|\Gamma^{(\nu)}\| \to \infty$ as $\nu \to \infty$. Replace $\Gamma^{(\nu)}$ with $\Gamma^{(\nu)}/\tau[\Gamma^{(\nu)}]$. Then $\tau[\Gamma^{(\nu)}] = 1$ and $\|\Gamma^{(\nu)}\| \to 0$. Hence we can find $\{\psi_k^{(\nu)}\}$ holomorphic on U_k so that $\|\psi^{(\nu)}\| < 2$ and $\Gamma_{jk}^{(\nu)} = \delta\{\psi_k^{(\nu)}\} = \psi_k^{(\nu)} - \psi_j^{(\nu)}$. By Vitali's theorem we can select a subsequence $\{\psi_j^{(\nu_\lambda)}\}$ such that $\psi_j^{(\nu_\lambda)}(z_j)$ converges uniformly on U_j^*. If $z \in U_k$ then z is in some U_j^*. Hence for each k, $\psi_k^{(\nu_\lambda)}(z) = \psi_j^{(\nu_\lambda)}(z) + \Gamma_{jk}^{(\nu_\lambda)}(z)$ converges uniformly on U_k so $\varphi_k^{(\nu_\lambda)}(z) \to \psi_k(z)$ uniformly on U_k. Since $\Gamma_{jk}^{(\nu_\lambda)} \to 0$, $\psi_j(z) = \psi_k(z)$ on $U_j \cap U_k$. Let $\hat{\psi}_j^{(\nu_\lambda)} = \psi_k^{(\nu_\lambda)} - \psi_j$. Then $\Gamma_{jk}^{(\nu_\lambda)} = \hat{\psi}_k^{(\nu_\lambda)} - \hat{\psi}_j^{(\nu_\lambda)}$. But $\|\hat{\psi}^{(\nu_\lambda)}\| < \frac{1}{2}$ for large ν, and thus $\tau[\Gamma^{(\nu_\lambda)}] < \frac{1}{2}$. This contradiction proves the lemma.   Q.E.D.

By Lemma 3.7 we can find K and $\{\varphi_{j|m+1}\}$ so that $\Gamma_{jk} = \varphi_{k|m+1} - \varphi_{j|m+1}$ and $\|\varphi_{m+1}\| \le K\|\Gamma\|$. Hence

$$\varphi_{j|m+1}(z_j)t^{m+1} \ll KK_1\left(2\kappa^2 n^2 + \frac{K_2}{\beta}\right)\frac{b}{c}\,A(t).$$

We can choose c so large that $KK_1(2\kappa^2 n^2 + K_2/\beta)(b/c) < 1$. Then $\varphi_{j|m+1}\,t^{m+1} \ll A(t)$ so $\varphi_j^{m+1}(z_j, t) - z_j \ll A(t)$. This completes the induction and proves that $\varphi_j(z_j, t) - z_j$ converges for small t, thus finishing the proof of the theorem.   Q.E.D.

There is the following:

THEOREM.   [Kodaira, Nirenberg, and Spencer, (1958)] Assume $H^2(M, \Theta) = 0$ (M is compact complex as always). Then for any element $\theta \in H^1(M, \Theta)$, there is a complex analytic family $\{M, |\,|t| < r,\ r > 0\}$, such that $M_0 = M$ and $(dM_t/dt)_{t=0} = \theta$.

PROBLEM.   Find an elementary proof of this. (In the analogous idea of proof the convergence gives trouble.)

We also have the completeness theorems.

DEFINITION 3.4.   We say that the family $(\mathcal{M}, B, \bar{\omega})$ is *complete* at $b \in B$ if, for any family $(\mathcal{N}, A, \pi)$ such that $\pi^{-1}(a) = \bar{\omega}^{-1}(b) = M_b$ there is a neighborhood $U \ni a$ and holomorphic maps $\Phi : \pi^{-1}(U) \to \mathcal{M}$, $h : U \to B$ such that $h(a) = b$,

$$\begin{array}{ccc}
\pi^{-1}(U) & \longrightarrow & \mathcal{M} \\
\downarrow{\scriptstyle \pi} & & \downarrow{\scriptstyle \bar{\omega}} \\
U & \xrightarrow{\ h\ } & B
\end{array} \qquad \text{commutes,}$$

Φ maps $\pi^{-1}(s)$ biholomorphically onto $\bar{\omega}^{-1}[h(s)]$ for each $s \in U$, and $\Phi : \pi^{-1}(a) = M_b \to M_b$ is the identity map.

Roughly, $(\mathcal{M}, B, \overline{\omega})$ is complete if it contains all sufficiently small deformations of M_b. The (holomorphic) tangent space T_b at b is the set

$$\left\{ \frac{\partial}{\partial t} \,\middle|\, \frac{\partial}{\partial t} = \sum_{v=1}^{m} c_v \left(\frac{\partial}{\partial t^v} \right) \right\}.$$

The map $\rho_b : (\partial/\partial t) \to (\partial M_t/\partial t)_{t=b} \in H^1(M_b, \Theta_b)$ defines a linear map $T_b \to H^1(M_b, \Theta_b)$.

THEOREM. [of completeness; see Kodaira and Spencer (1958b)] If $\rho_b(T_b) = H^1(M_b, \Theta_b)$, then $(\mathcal{M}, B, \overline{\omega})$ is complete at b.

REMARKS. (1) Theorem 3.2 is a corollary of this theorem. (2) This theorem of completeness can be proved by the same elementary method used to prove Theorem 3.2.

4. Exact Sequences

As usual X is a paracompact Hausdorff space, $\mathcal{S}$ is a sheaf over X and $\overline{\omega} : \mathcal{S} \to X$ is its projection map.

DEFINITION 4.1. $\mathcal{S}' \subseteq \mathcal{S}$ is a *subsheaf* if

(1) $\mathcal{S}'$ is open,
(2) $\overline{\omega}(\mathcal{S}') = X$,
(3) $\overline{\omega}^{-1}(x) \cap \mathcal{S}' = \mathcal{S}'_x$ is an R-submodule of $\mathcal{S}_x$.

Let $\mathcal{S}''$ be a sheaf over X with projection $\overline{\omega}''$.

DEFINITION 4.2. A *homomorphism h* of $\mathcal{S}$ into $\mathcal{S}''$ is a continuous map of $\mathcal{S}$ into $\mathcal{S}''$ such that

(1) $\overline{\omega}'' \circ h = \overline{\omega}$,
(2) $h : \mathcal{S}_x \to \mathcal{S}''_x$ is an R-homomorphism.

REMARK. h is a local homeomorphism. We define the kernel of h to be
$$\ker h = \{ s \mid h(s) = 0 \} \text{ where } "t = 0" \text{ means } t_x = 0_x \in \mathcal{S}_x.$$

LEMMA 4.1. $\ker h$ is a subsheaf of $\mathcal{S}$.

LEMMA 4.2. $h(\mathcal{S})$ is a subsheaf of $\mathcal{S}''$.
The proofs are left to the reader.

Let $\mathscr{S}' \in \mathscr{S}$ be a subsheaf. Let $Q_x = \mathscr{S}_x/\mathscr{S}'_x$ which is an R-module. Define $h_x : \mathscr{S}_x \to Q_x$ to be the natural homomorphism. Let $Q = \bigcup_{x \in X} Q_x$. Define π on Q by $\pi(Q_x) = x$; h is defined $h : \mathscr{S} \to Q$ by $h(s) = h_x(s)$ for $s \in \mathscr{S}_x$. We give Q a topology by saying $\mathscr{U}$ is open if and only if $h^{-1}(\mathscr{U})$ is open in $\mathscr{S}$ (the quotient topology).

LEMMA 4.3. Q is a sheaf and $h : \mathscr{S} \to Q$ is a surjective homomorphism.

Proof.   Left to the reader.

DEFINITION 4.3. Q is the *quotient sheaf* of $\mathscr{S}$ by $\mathscr{S}'$ and we write $Q = \mathscr{S}/\mathscr{S}'$.

A sequence

$$\mathscr{S}_0 \xrightarrow{h_0} \mathscr{S}_1 \xrightarrow{h_1} \mathscr{S}_2 \longrightarrow \cdots \longrightarrow \mathscr{S}_n \xrightarrow{h_n} \mathscr{S}_{n+1} \xrightarrow{h_{n+1}} \cdots$$

of sheaves is *exact* if $h_v(\mathscr{S}_v) = \ker(h_{v+1})$ for all v.

Suppose given sheaves $\mathscr{S}$ and $\mathscr{S}''$ over X and a homomorphism $h : \mathscr{S} \to \mathscr{S}''$. Let U be an open subset of X and $\Gamma(U, \mathscr{S})$ be the set of all sections of $\mathscr{S}$ over U, Then $h \circ \sigma \in \Gamma(U, \mathscr{S}'')$ if $\sigma \in \Gamma(U, \mathscr{S})$ so h induces a map $h : \Gamma(U, \mathscr{S}) \to \Gamma(U, \mathscr{S}'')$. Let $\mathscr{U} = \{U_j\}$ be a locally finite covering and $C^q(\mathscr{U}, \mathscr{S})$ be the space of q-cochains $c^q = \{\sigma_{j_0 \cdots j_q}\}$ where $\sigma_{j_0 \cdots j_q} \in \Gamma(U_{j_0} \cap \cdots \cap U_{j_q}, \mathscr{S})$. Then h induces a map $h : C^q(\mathscr{U}, \mathscr{S}) \to C^q(\mathscr{U}, \mathscr{S}'')$ defined by $h\, c^q = \{h\sigma_{j_0 \cdots j_q}\}$ Then we have:

LEMMA 4.4. $h \circ \delta = \delta \circ h$.

Proof.   Obvious.

Hence h maps $Z^q(\mathscr{U}, \mathscr{S})$ into $Z^q(\mathscr{U}, \mathscr{S}'')$ and thus h induces a homomorphism $h : H^q(\mathscr{U}, \mathscr{S}) \to H^q(\mathscr{U}, \mathscr{S}'')$. Let $\mathscr{W} = \{\mathscr{W}_\lambda\}$ be a refinement of $\mathscr{U}$, $\mathscr{W} > \mathscr{U}$. Then

$$
\begin{array}{ccc}
H^q(\mathscr{U}, \mathscr{S}) & \xrightarrow{\ h\ } & H^q(\mathscr{U}, \mathscr{S}'') \\
\downarrow{\scriptstyle \Pi^{\mathscr{U}}_{\mathscr{W}}} & & \downarrow{\scriptstyle \Pi^{\mathscr{U}}_{\mathscr{W}}} \\
H^q(\mathscr{W}, \mathscr{S}) & \xrightarrow{\ h\ } & H^q(\mathscr{W}, \mathscr{S}'')
\end{array}
$$

commutes. Hence h induces a homomorphism $h : H^q(X, \mathscr{S}) \to H^q(X, \mathscr{S}'')$.

THEOREM 4.1. Assume that

$$0 \to \mathscr{S}' \xrightarrow{i} \mathscr{S} \xrightarrow{h} \mathscr{S}'' \to 0$$

is exact. Then there is a homomorphism δ^* such that

$$0 \longrightarrow H^0(X, \mathscr{S}') \xrightarrow{\;i\;} H^0(X, \mathscr{S}) \xrightarrow{\;h\;} H^0(X, \mathscr{S}'')$$
$$\xrightarrow{\;\delta^*\;} H^1(X, \mathscr{S}') \xrightarrow{\;i\;} H^1(X, \mathscr{S}) \longrightarrow \cdots$$

is exact.

Proof. i is injective so $\mathscr{S}' \cong i(\mathscr{S}') \subset \mathscr{S}$, and $i(\mathscr{S}') = \ker h$. Thus we consider $\mathscr{S}' \subset \mathscr{S}$ where $\mathscr{S}' = \ker h$ and i is the inclusion map. Recall that $H^0(X, \mathscr{S}) = Z^0(X, \mathscr{S}) = \Gamma(X, \mathscr{S})$. Since $\Gamma(X, \mathscr{S}') \subset \Gamma(X, \mathscr{S})$, we see that $0 \to H^0(X, \mathscr{S}') \xrightarrow{i} H^0(X, \mathscr{S})$ is exact. If $\sigma \in \Gamma(X, \mathscr{S})$, then $h\sigma = 0$ if and only if $\sigma \in \Gamma(X, \mathscr{S}')$; so $H^0(X, \mathscr{S}') \xrightarrow{i} H^0(X, \mathscr{S}) \xrightarrow{h} H^0(X, \mathscr{S}'')$ is exact.

LEMMA 4.5. $H^0(X, \mathscr{S}) \xrightarrow{h} H^0(X, \mathscr{S}'') \xrightarrow{\;\delta^*\;} H^1(X, \mathscr{S}')$ is exact (where we must define δ^*).

Proof. Let $\sigma'' \in \Gamma(X, \mathscr{S}'')$. Since h is a local homomorphism there is a section $\tau_y \in \Gamma(U_y, \mathscr{S})$ over a small neighborhood U_y of y such that $h\,\tau_y(x) = \sigma''(x)$ for $x \in U_y$. Now $\{U_y \mid y \in X\}$ covers X and we have a locally finite refinement $\mathscr{U} = \{U_j\}$ of $\{U_y\}$, that is, there is a map $j \to y(j)$ such that $U_j \subseteq U_{y(j)}$. Set $\tau_j = r_{U_j}\, \tau_{y(j)} \in \Gamma(U_j, \mathscr{S})$. Then $h\,\tau_j = \sigma''$ where defined. Let $c^0 = \{\tau_j\} \in c^0(\mathscr{U}, \mathscr{S})$. Then:

DEFINITION $(4.4)_1$. $\delta^*\sigma'' = [\delta c^0] \in H^1(X, \mathscr{S}')$ where for any $c^q \in Z^q(\mathscr{U}, \mathscr{S})$, $[c^q]$ denotes the cohomology class in $H^q(X, \mathscr{S})$ of c^q. (One should check that δ^* is well defined.)

Since $\delta c^0 = \{\tau_k - \tau_j\}$ and $h\tau_k - h\tau_j = \sigma'' - \sigma'' = 0$, we see that $\delta c^0 \in Z^1(\mathscr{U}, \mathscr{S}')$ so Definition $(4.4)_1$ makes sense. Exactness means $\delta^*\sigma'' = 0$ if and only if $\sigma'' = h\sigma$ for some $\sigma \in \Gamma(X, \mathscr{S})$. So suppose $\delta^*\sigma'' = [\delta c^0] = 0$. Then $\delta c^0 = \delta c^{0'}$ where $c^{0'} = \{\tau_j'\} \in c^0(\mathscr{U}, \mathscr{S}')$. So $c^0 - c^{0'} = \sigma \in Z^0(X, \mathscr{S}) = \Gamma(X\ \mathscr{S})$, and $h\sigma = hc^0 = \{h\tau_j\} = \sigma''$. Now suppose $\sigma'' = h\sigma$. Then $h(\tau_j - \sigma) = 0$, so $\tau_j - \sigma \in \Gamma(U_j, \mathscr{S}')$. Set $c_0' = \{\tau_j - \sigma\} = c_0 - \sigma \in C^0(\mathscr{U}, \mathscr{S}')$. Then $\delta c_0' = \delta c_0$ since $\sigma \in Z^0(X, \mathscr{S})$ and hence $\delta^*\sigma'' = [\delta c_0] = [\delta c_0'] = 0$.   Q.E.D.

We now turn to check that

$$H^0(X, \mathscr{S}'') \xrightarrow{\;\delta^*\;} H^1(X, \mathscr{S}') \xrightarrow{\;i\;} H^1(X, \mathscr{S})$$

is exact. Take $c^{1'} \in Z^1(\mathcal{U}, \mathcal{S}')$. If $[c^{1'}] = \delta*\sigma'' = [\delta c^0]$, then $i[c^{1'}] = 0$, and if $i[c^{1'}] = 0$, then $c^{1'} = \delta c^0$; so $0 = hc^{1'} = h\delta c^0 = \delta hc^0$. Thus hc^0 defines an element $\sigma'' \in \Gamma(X, \mathcal{S}'')$. By definition, $[c^{1'}] = \delta*\sigma''$.

We want to prove exactness

$$\longrightarrow H^q(X, \mathcal{S}') \xrightarrow{\ i\ } H^q(X, \mathcal{S}) \xrightarrow{\ h\ }$$

$$H^q(X, \mathcal{S}'') \xrightarrow{\ \delta*\ } H^{q+1}(X, \mathcal{S}') \longrightarrow .$$

LEMMA 4.6. Given $c^{q''} \in C^q(\mathcal{U}, \mathcal{S}'')$, then we can find a locally finite refinement $\mathcal{W}$ and $c^q \in C^q(\mathcal{W}, \mathcal{S})$ so that $\Pi^{\mathcal{U}}_{\mathcal{W}} c^{q''} = hc^q$.

Proof. We give proof for $q = 2$. Let $\mathcal{U} = \{U_j\}$, $c^{q''} = \{\sigma''_{ijk}\}$, where $\sigma''_{ijk} \in \Gamma(U_i \cap U_j \cap U_k, \mathcal{S}'')$. Choose a covering $\mathcal{V} = \{V_j\}$ such that $\overline{V}_j \subset U_j$. Since $\mathcal{U}$ is locally finite, a given $y \in X$ belongs to only finitely many U_j. We choose a neighborhood N_y of y sufficiently small so that

(1) if $y \in U_k \cap U_j \cap U_k$ there is $\tau \in \Gamma(N_y, \mathcal{S})$ with $\sigma''_{ijk}(x) = h\tau(x)$ for $x \in N_y$ (remember h is a local homeomorphism),

(2) for each y there is V_j such that $N_y \subset V_j$,

(3) if $N_y \cap \overline{V}_j \neq \phi$ then $N_y \subset U_j$.

Then $\{N_y \mid y \in X\}$ covers X and we can choose $\mathcal{W} = \{W_\lambda\}$ a locally finite refinement of $\{N_y\}$. Hence there is a map $\lambda \to y_\lambda$ such that $W_\lambda \subset N_{y_\lambda}$. By (2) $N_{y_\lambda} \subset V_{j_\lambda}$, so $\mathcal{W} > \mathcal{V} > \mathcal{U}$. Define $\tau = \{\tau_{\lambda\mu\nu}\} \in C^2(\mathcal{W}, \mathcal{S})$ as follows: we have $W_\lambda \subset V_i$, $W_\mu \subset V_j$, $W_\nu \subset C_k$ where $i = j_\lambda$, $j = j_\lambda$, $k = j_\nu$. By (3) if $N_{y_\lambda} \cap V_j \neq \phi$, $N_{y_\lambda} \subset U_j$, and $N_{y_\lambda} \cap V_k \neq \phi$ gives $N_{y_\lambda} \subset U_k$, and so on. We are assuming $W_\lambda \cap W_\mu \cap W_\nu \neq \phi$, and $W_\lambda \subset N_{y_\lambda} \subset V_i$, and so on. Hence it follows that $y_\lambda \in N_{y_\lambda} \subset U_k \cap U_j \cap U_k$. By (1) $\sigma''_{ijk}(x) = h\tau(x)$ for $x \in N_{y_\lambda}$ where $\tau \in \Gamma(N_{y_\sigma}, \mathcal{S})$. Let $\tau_{\lambda\mu\nu} = r_{W_\lambda \cap W_\mu \cap W_\nu}(\tau)$. Then

$$h\tau_{\lambda\mu\nu} = r_{W_\lambda \cap W_\mu \cap W_\nu} \sigma''_{j_\lambda j_\mu j_\nu}.$$

Let $c^2 = \{\tau_{\lambda\mu\nu}\}$. Then $hc^2 = \Pi^{\mathcal{U}}_{\mathcal{W}} c^{2''}$.   Q.E.D.

Let us prove that

$$H^q(X, \mathcal{S}) \xrightarrow{\ i\ } H^q(X, \mathcal{S}) \xrightarrow{\ h\ } H^q(\mathcal{S}'')$$

is exact. $hi = 0$ is clear. Suppose $\eta \in H^q(X, \mathcal{S})$ and $h\eta = 0$. Then $\eta = [c^q]$, $c^q \in Z^q(\mathcal{U}, \mathcal{S})$ for some $\mathcal{U}$ and $\Pi^{\mathcal{U}}_{\mathcal{V}} hc^q = \delta c^{q-1''}$ for some $\mathcal{V}$ and $c^{q-1''} \in C^{q-1}(\mathcal{V}, \mathcal{S}'')$. By the lemma $\Pi^{\mathcal{V}}_{\mathcal{W}} c^{q-1''} = h\, t^{q-1}$ for some $\mathcal{W}$ and $t^{q-1} \in C^{q-1}(\mathcal{W}, \mathcal{S})$. Thus

$$h\Pi^{\mathcal{U}}_{\mathcal{W}} c^q - h\, \delta t^{q-1} = 0, \qquad \text{so } \Pi^{\mathcal{U}}_{\mathcal{W}} c^q - \delta t^{q-1} = c^{q'}$$

where $c^{q'} \in Z^q(\mathcal{W}, \mathcal{S}') \subseteq C^q(\mathcal{W}, \mathcal{S})$. Finally we get $\eta = [c^q] = [c^{q'}]$ so $\eta = i\eta'$ where $\eta' = [c^{q'}] \in H^q(X, \mathcal{S}')$.

Next we prove that

$$H^q(X, \mathscr{S}) \xrightarrow{\;h\;} H^q(X, \mathscr{S}'') \xrightarrow{\;\delta^*\;} H^{q+1}(X, \mathscr{S}')$$

is exact. We must define δ^*. Take $\eta'' \in H^q(X, \mathscr{S}'')$. Then $\eta'' = [c^{q''}]$, $c^{q''} \in Z^q(\mathscr{U}, \mathscr{S}'')$. By the lemma there is $t^q \in H^q(\mathscr{W}, \mathscr{S})$ such that $h\, t^q = \Pi_{\mathscr{W}}^{\mathscr{U}}\, c^{q''}$.

DEFINITION 4.4. $\delta^* \eta'' = [\delta t^q] \in H^{q+1}(X, \mathscr{S}')$. Again we should check that δ^* is well defined. For the moment denote $\Pi_{\mathscr{W}}^{\mathscr{U}}$ by Π. Suppose $\delta^* \eta = 0$. Then $\Pi \delta t^q = \delta b^{q'}$ for some $b^{q'} \in C^q(\mathscr{W}, \mathscr{S}')$. Thus $\delta(\Pi t^q - b^{q'}) = 0$ and $\Pi\, t^q - b^q = c^q \in Z^q(\mathscr{W}, \mathscr{S})$. So let $\eta = [c^q] \in H^q(X, \mathscr{S})$. Then $h\eta = [hc^q] = [\Pi\, ht^q] = [\Pi\, c^{q''}] = \eta''$. Suppose conversely that $\eta'' = h\eta$. Let $\eta = [c^q]$. Then $n'' = [hc^q]$ and $ht^q = \Pi\, hc^q$ by definition of t^q. So $t^q - \Pi c^q = a^{q'} \in C^q(\mathscr{W}, \mathscr{S}')$ and $\delta^* \eta'' = [\delta t^q] = [\delta a^{q'}]$. Thus $\delta^* \eta'' = 0$.

Finally we prove that

$$H^q(X, \mathscr{S}'') \xrightarrow{\;\delta^*\;} H^{q+1}(X, \mathscr{S}') \xrightarrow{\;i\;} H^{q+1}(X, \mathscr{S})$$

is exact. Certainly $i\, \delta^* \eta'' = 0$. Since by definition $\delta^* \eta = [\delta t^q]$ where $t^q \in C^q$ $(\mathscr{W}, \mathscr{S})$ so $i\delta^* \eta = 0$. Suppose $i\eta' = 0$, $\eta' = [c^{q+1'}] \in H^{q+1}(X, \mathscr{S}')$. Then $\Pi c^{q+1'} = \delta t^q$ for $t^q \in C^q(\mathscr{W}, \mathscr{S})$. Then $0 = \delta\ ht^q \in Z^q(\mathscr{W}, \mathscr{S}'')$ and $\eta'' = [ht^q] \in H^q(X, \mathscr{S}'')$. Since $\delta^* \eta'' = [\delta t^q] = \eta'$ we are finished.   Q.E.D.

[If the reader wishes more details for these elementary properties of sheaves he may consult Hirzebruch (1962).]

Next we prove functoriality.

THEOREM 4.2. If

$$
\begin{array}{ccccccccc}
0 & \longrightarrow & \mathscr{S}' & \xrightarrow{\;i\;} & \mathscr{S} & \xrightarrow{\;h\;} & \mathscr{S}'' & \longrightarrow & 0 \\
 & & \downarrow{\scriptstyle \varphi'} & & \downarrow{\scriptstyle \varphi} & & \downarrow{\scriptstyle \varphi''} & & \\
0 & \longrightarrow & \mathscr{T}' & \xrightarrow{\;\tau\;} & \mathscr{T} & \xrightarrow{\;k\;} & \mathscr{T}'' & \longrightarrow & 0
\end{array}
$$

is exact and commutative, then

$$
\begin{array}{ccccccccccc}
0 & \longrightarrow & H^0(X, \mathscr{S}') & \xrightarrow{\;i\;} & H^0(X, \mathscr{S}) & \xrightarrow{\;h\;} & H^0(X, \mathscr{S}'') & \xrightarrow{\;\delta^*\;} & H^1(X, \mathscr{S}') & \longrightarrow & \cdots \\
 & & \downarrow{\scriptstyle \varphi'} & & \downarrow{\scriptstyle \varphi} & & \downarrow{\scriptstyle \varphi''} & & \downarrow{\scriptstyle \varphi'} & & \\
0 & \longrightarrow & H^0(X, \mathscr{T}) & \xrightarrow{\;\tau\;} & H^0(X, \mathscr{T}) & \longrightarrow & H^0(X, \mathscr{T}'') & \xrightarrow{\;\delta^*\;} & H^1(X, \mathscr{T}) & \longrightarrow & \cdots
\end{array}
$$

is exact and commutative.

Proof. We need only prove commutativity. We check that

$$
\begin{array}{ccc}
H^q(X, \mathscr{S}'') & \xrightarrow{\ \delta^*\ } & H^{q+1}(X, \mathscr{S}') \\
\Big\downarrow{\varphi''} & & \Big\downarrow{\varphi'} \\
H^q(X, \mathscr{T}'') & \xrightarrow{\ \delta^*\ } & H^{q+1}(X, \mathscr{T}')
\end{array}
$$

commutes. The rest is easy. Let $\eta'' \in H^q(X, \mathscr{S}'')$. Then there is $c^{q''} \in Z^q(\mathscr{W}, \mathscr{S}'')$ and $t^q \in C^q(\mathscr{W}, \mathscr{S})$ such that $\eta'' = [c^{q''}]$ and $c^{q''} = ht^q$. Thus $\delta^* \eta'' = [\delta t^q]$. However, $\varphi' \delta^* \eta'' = [\varphi' \delta t^q] = [\delta \varphi t^q]$, and $\varphi'' \eta'' = [\varphi'' ht^q] = [\varphi'' ht^q] = [h \, \varphi t^q]$. Thus $\delta^* \varphi'' \eta'' = [\delta \, \varphi t^q] = \varphi' \delta^* \eta''$.   Q.E.D.

We give a brief discussion of fine sheaves.

DEFINITION 4.5. $\mathscr{S}$ is a *fine sheaf* if for any locally finite covering $\{U_j\}$ of X there exists a set $\{h_j\}$ of homomorphisms $h_j \colon \mathscr{S} \to \mathscr{S}$ such that

(1) $h_j \mathscr{S}_x = 0$ for $x \notin \overline{W}_j$, where $\overline{W}_j \subseteq U_j$ is a closed subset of U_j,
(2) $\sum_j h_j = id.$

EXAMPLE.   Let $\mathscr{D}$ be the sheaf of germs of differentiable functons on a differentiable manifold X. We have a *partition of unity* subordinate to U_j; that is, a set $\{\rho_j\}$ of differentiable functions $\rho_j = \rho_j(x)$ on X such that

(1) $\rho_j(x) = 0$ for $x \notin \overline{W}_j$,
(2) $\sum \rho_j = 1.$

For any local differentiable function $f = f(x)$ on X, define $h_j f = \rho_j(x) f(x)$. Then h_j induces a homomorphism $h_j \colon \mathscr{D} \to \mathscr{D}$. Using these $\{h_j\}$ we see that $\mathscr{D}$ is fine.

THEOREM 4.3.   If $\mathscr{S}$ is a fine, then $H^q(X, \mathscr{S}) = 0$ for $q \geq 1$.

Proof.   We give the proof for the case $q = 2$. Let c^2 be a cocycle, $c^2 = \{\sigma_{ijk}\} \in Z^2(\mathscr{U}, \mathscr{S})$, with $\sigma_{ijk} \in \Gamma(U_i \cap U_j \cap U_k, \mathscr{S})$. By fine-ness we have the $\{h_j\}$ in Definition 4.5. Since $\delta c^2 = 0$, if $U_i \cap U_j \cap U_k \cap U_\ell \neq \phi$ then $\sigma_{jk\ell} - \sigma_{ik\ell} + \sigma_{ij\ell} - \sigma_{ijk} = 0$. Since $h_i \sigma_{ijk}(x) = 0$ for $x \notin \overline{W}_i$, $h_i \sigma_{ijk}$ can be extended to $\tau_{ijk} \in \Gamma(U_j \cap U_k, \mathscr{S})$ by setting $\tau_{ijk}(x) = 0$ for $x \in U_j \cap U_k - \overline{W}_i$. For fixed (j, k) we have only a finite number of U_i with $U_i \cap U_j \cap U_k \neq \phi$ and for each i we have τ_{ijk} from $h_i \sigma_{ijk}$. We set $\tau_{jk} = \sum_i \tau_{ijk}$. Then

$$
h_i \sigma_{jk\ell} = h_i \sigma_{ik\ell} - h_i \sigma_{ij\ell} + h_i \sigma_{ijk}.
$$

Thus $\sigma_{jk\ell} = \tau_{k\ell} - \tau_{j\ell} + \tau_{jk}$ so $c^2 = \delta c^1$ where $c^1 = \{\tau_{jk}\}$. It is easy to give this proof for any $q \geq 1$.   Q.E.D.

5.   Vector Bundles

We give a brief review of vector bundles. Again, a good reference for this section is Hirzebruch (1962). Let M be a complex (differentiable) manifold.

DEFINITION 5.1. By a *complex analytic (differentiable) vector bundle* ($\mathbb{C}^n$ or $\mathbb{R}^n$ bundle) we mean a complex (differentiable) manifold F together with a holomorphic (differentiable) map $\pi : F \to M$ onto M such that, for a sufficiently fine locally finite covering $\mathcal{U} = \{U_j\}$ of M:

(1)   There is an analytic (differentiable) equivalence f_j between $\pi^{-1}(U_j)$ and $U_j \times \mathbb{C}^n$ (or $U_j \times \mathbb{R}^n$) such that

$$
\begin{array}{ccc}
\pi^{-1}(U_j) & \xrightarrow{\;f_j\;} & U_j \times \mathbb{C}^n \\
{\scriptstyle \pi}\downarrow & & \downarrow{\scriptstyle \pi_j} \\
U_j & \xrightarrow{\;id\;} & U_j
\end{array}
$$

commutes, where $\pi_j(z_j, \zeta) = z_j$.

(2)   If $(z, \zeta_j^1, \cdots, \zeta_j^n) \in U_j \times \mathbb{C}^n$ (or $U_j \times \mathbb{R}^n$) and $(z, \zeta_k^1, \cdots, \zeta_k^n) \in U_k \times \mathbb{C}^n$ (or $U_j \times \mathbb{R}^n$), then

$$
f_j \circ f_k^{-1}(z, \zeta_k^1, \cdots, \zeta_k^n) = \sum_{\beta=1}^{n} f_{jk\beta}^{\alpha}(z)\zeta_k^{\beta},
$$

where $f_{jk\beta}^{\alpha}(z)$ are holomorphic (differentiable) functions on $U_j \cap U_k$. In vector notation f_{jk} is the matrix $(f_{jk\beta}^{\alpha})$ and $\zeta_j = (\zeta_j^1, \cdots, \zeta_j^n)$, and then

$$
\zeta_j = f_{jk}(z)\,\zeta_k \qquad \text{for } z \in U_j \cap U_k.
$$

We call $\pi^{-1}(z)$ the *fibre* of F over. By (1) and (2) we can give it a vector space structure $\pi^{-1}(z) = z \times \mathbb{C}^n$ [or $\pi^{-1}(z) = z \times \mathbb{R}^n$].

DEFINITION 5.2. We say that F and F' are holomorphically (or differentiably) equivalent if there is a biholomorphic (bidifferentiable) map $\varphi : F \to F'$ such that

(1) 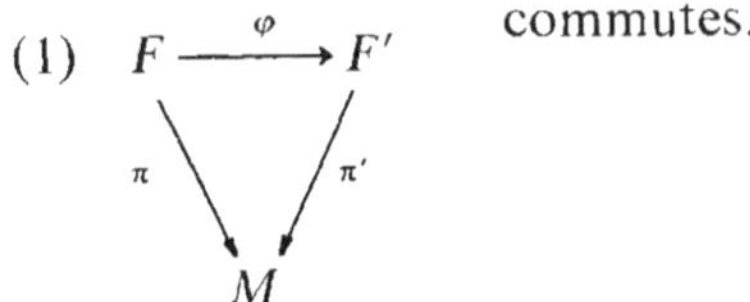   commutes.

(2)   On each fibre φ is a linear transformation, that is, if U_j is chosen as in Definition 5.1, then there is a holomorphic (differentiable) matrix-valued function h_j on U_j such that $f_j' \circ \varphi \circ f_j^{-1} = h_j$, that is, $\zeta_j'^{\alpha} = \sum_{\beta} h_{j\beta}^{\alpha}(z)\,\zeta_j^{\beta}$.

In particular, if F is a $\mathbb{C}^1$ bundle and F is trivial over U_j (that is, $\pi^{-1}(U_j) = U_j \times \mathbb{C}$), then $z \in U_j \cap U_k$ implies that (z, ζ_j) is identified with (z, ζ_k) if and only if $\zeta_j = f_{jk}(z)\zeta_k$ where $f_{jk}(z)$ is a nonvanishing holomorphic (differentiable function) of $z \in U_j \cap U_k$. Let $\mathcal{O}^*$ (or $\mathcal{D}^*$) be the sheaf over M of nonvanishing holomorphic (or differentiable) functions in which the module operation on each stalk is multiplication and the ring $R = \mathbb{Z}$ so here instead of $\alpha f(z) + \beta g(z)$ we have $[f(z)]^{\alpha}[g(z)]^{\beta}$, $\alpha, \beta \in \mathbb{Z}$. Consider f_{jk} as an element of

$$\Gamma\left(U_j \cap U_k, \left\{{\mathcal{O}^* \atop \mathcal{D}^*}\right\}\right).$$

On $U_i \cap U_j \cap U_k$ we have

$$f_{ik}(z) = f_{ij}(z)f_{jk}(z).$$

Hence $\{f_{jk}\} \in Z^1(\mathcal{U}, \mathcal{O}^* \text{ (or } \mathcal{D}^*))$ and two bundles F and F' are equivalent if there are nonvanishing functions $h_j(z)$ on U_j such that $f_j'^{-1} \circ h_j \circ f_j = f_k'^{-1} \circ h_k \circ f_k$ on $U_j \cap U_k$. This is equivalent to

$$f_{jk}'(z) = h_j(z)f_{jk}(z)h_k^{-1}(z)$$

or $\{f_{jk}'\} = \{f_{jk}\} \cdot \delta\{h_v^{-1}\}$, that is, $\{f_{jk}'\}$ is cohomologous to $\{f_{jk}\}$. Thus an equivalence class of bundles defines an element $[\{f_{jk}\}] \in H^1(\mathcal{U}, \mathcal{O}^*(\text{or } \mathcal{D}^*)) \subseteq H^1(M, \mathcal{O}^*(\text{or } \mathcal{D}^*))$. Conversely it is easy to construct a bundle from an element of $H^1(M, \mathcal{O}^*(\text{or } \mathcal{D}^*))$. Thus we have a 1-1 correspondence between equivalence classes of $\mathbb{C}^1$ bundles and classes in $H^1(M, \mathcal{O}^*(\text{or } \mathcal{D}^*))$. We shall always identify equivalent $\mathbb{C}^1$ bundles and we call $H^1(M, \mathcal{O}^*(\text{or } \mathcal{D}^*))$ the *group of $\mathbb{C}^1$ bundles* over M. It has a natural group structure if we define $F \cdot G = \{f_{jk}g_{jk}\}$ where $F = \{f_{jk}\}$ and $G = \{g_{jk}\}$.

We construct an important invariant of $\mathbb{C}^1$ bundles. For any germ of a holomorphic function f, $e^{2\pi i f} \in \mathcal{O}^*$. Thus we get an exact sequence

$$0 \to \mathbb{Z} \to \mathcal{O} \to \mathcal{O}^* \to 0,$$

where $\mathbb{Z}$ is the sheaf of germs of locally constant integer valued functions on M. We also have the following commuting, exact diagram:

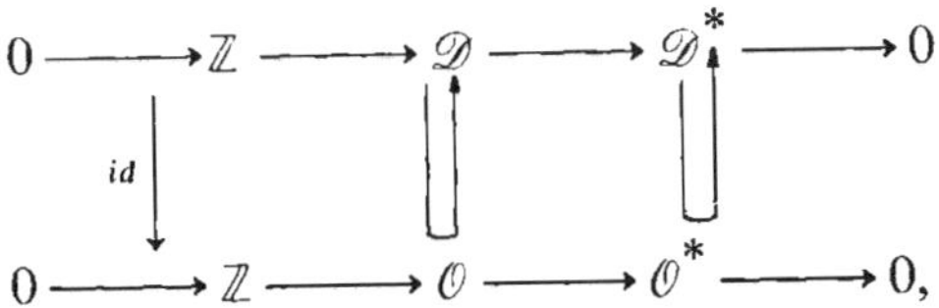

since $\mathcal{O} \subset \mathcal{D}$ and $\mathcal{O}^* \subset \mathcal{D}^*$. This yields the exact commutative sequences

$$\cdots \longrightarrow H^1(M, \mathbb{Z}) \longrightarrow H^1(M, \mathcal{O}) \xrightarrow{\ \delta\ } H^1(M, \mathcal{O}^*)$$

$$\xrightarrow{\ \delta^*\ } H^2(M, \mathbb{Z}) \longrightarrow H^2(M, \mathcal{O}) \longrightarrow \cdots$$

$$\cdots \longrightarrow H^1(M, \mathbb{Z}) \longrightarrow H^1(M, \mathcal{D}) \longrightarrow H^1(M, \mathcal{D}^*)$$

$$\xrightarrow{\ \delta^*\ } H^2(M, \mathbb{Z}) \longrightarrow H^2(M, \mathcal{D}) \longrightarrow \cdots.$$

Now $\mathcal{D}$ is a fine sheaf, so $H^q(M, \mathcal{D}) = 0$ for $q \geq 1$. Thus δ^* is an isomorphism $\delta^* : H^1(M, \mathcal{D}^*) \to H^2(M, \mathbb{Z})$.

DEFINITION 5.3. $c(F) = \delta^*(F)$ is the (first) *Chern class* of F.

We remark that $c(F) = c(G)$ if and only if $\varphi(F)$ and $\varphi(G)$ are equal in $H^1(M, \mathcal{D}^*)$ where $\varphi : H^1(M, \mathcal{O}^*) \to H^1(M, \mathcal{D}^*)$ is induced by $\mathcal{O}^* \subset \mathcal{D}^*$, where $F, G \in H^1(M, \mathcal{O}^*)$. Hence F and G are *differentiably* equivalent; that is, there are nonvanishing differentiable functions h_j such that $\{f_{jk}\} = \{h_j\, g_{jk}\, h_k^{-1}\}$ where $F = \{f_{jk}\}$, $G = \{g_{jk}\}$. Thus:

PROPOSITION 5.1.   The Chern class $c(F)$ of a complex analytic $\mathbb{C}^1$ bundle represents the differentiable equivalence class of F.

Let us give an explicit description of $c(F)$. If $F = \{f_{jk}\}$, $f_{ij} \cdot f_{jk} \cdot f_{ki} = 1$ and

$$\log f_{ij} + \log f_{jk} + \log f_{ki} = 2\pi i\, c_{ijk},$$

where $i = \sqrt{-1}$. Then $c(F) = \{c_{ijk}\} \in H^2(M, \mathbb{Z})$.

Next consider $\mathbb{C}^n$ bundles. Let $n \geq z$ and F be a $\mathbb{C}^n$ bundle defined by $\{f_{jk}\}$. Let $\mathfrak{G}$ be the sheaf over M of germs of matrix-valued holomorphic functions $f(z) = f_\beta^\alpha(z)$ with $\det f_\beta^\alpha(z) \neq 0$ where the module operation is matrix multiplication. Note that the operation is *not* commutative. We cannot define the higher cohomology groups of $\mathfrak{G}$ but we can define the following objects: Let $\mathcal{U} = \{U_j\}$ be a locally finite open covering of X. A 0-*cochain* $c^0 = \{f_j\}$, $f_j \in \Gamma(U_j, \mathfrak{G})$ is a set of sections of $\mathfrak{G}$ over U_j; a 1-cochain $c^1 = \{f_{jk}\}$ where $f_{jk} \in \Gamma(U_{jk}, \mathfrak{G})$; a 1-cocycle c^1 is a 1-cochain such that $f_{jk}(z) = f_{ij}(z) \cdot f_{jk}(z)$ for $z \in U_i \cap U_j \cap U_k$. Let $Z^1(\mathcal{U}, \mathfrak{G})$ be the set of all 1-cocycles. We note that $Z^1(\mathcal{U}, \mathfrak{G})$ is *not* a group.

DEFINITION 5.4.   We say that $\{f_{ik}\}$ and $\{g_{ik}\} \in Z^1(\mathcal{U}, \mathfrak{G})$ are *equivalent* if there exists $\{h_j\}$ such that $g_{ik} = h_i f_{ik} h_k^{-1}$. Let $H^1(\mathcal{U}, \mathfrak{G})$ be the set of equivalence classes of 1-cocycles. We define $H^1(M, \mathfrak{G}) = \lim_{\mathcal{U}} H^1(\mathcal{U}, \mathfrak{G})$.

PROPOSITION 5.2.   Each element of $H^1(M, \mathfrak{G})$ represents an equivalence class of complex analytic $\mathbb{C}^n$ bundles.

Proof.   Left to the reader.

We now list some methods of forming new bundles from old bundles. Let F be a $\mathbb{C}^n$ bundle, G be a $\mathbb{C}^m$ bundle, $\mathcal{U} = \{U_j\}$ a trivializing covering and $F = \{f_{jk}\}$, $G = \{g_{jk}\}$. We define the following new objects:

(1)   *Whitney Sum* $F \oplus G$.   This is a $\mathbb{C}^{n+m}$ bundle which is defined by the cocycle $\{h_{jk}\}$ where

$$h_{jk}(z) = \begin{pmatrix} f_{jk} & 0 \\ 0 & g_{jk} \end{pmatrix}.$$

(2)   *Tensor Product* $F \otimes G$.   This is a $\mathbb{C}^{nm}$ bundle defined by $\{h_{jk}\}$ where $h_{jk}(z) = f_{jk}(z) \otimes g_{jk}(z)$. Recall that

$$h_{jk} = \begin{pmatrix} h_{11}^{11} & h_{21}^{11} & \cdots & h_{n1}^{11} & \cdots & h_{nm}^{11} \\ h_{11}^{21} & \cdots & & & & \\ h_{11}^{nm} & \cdots & & & & h_{nm}^{nm} \end{pmatrix},$$

and here $h_{jk\beta\mu}^{\alpha\lambda} = f_{jk\beta}^{\alpha} g_{jk\mu}^{\lambda}$. A point in $F \otimes G$ has coordinates $(z, \zeta_j^{11}, \cdots, \zeta_j^{1m}, \cdots, \zeta_j^{nm})$, where (z, ζ_j) and (z, ζ_k) are identified for $z \in U_j \cap U_k$ if and only if

$$\zeta_j^{\alpha\lambda} - \sum f_{jk\beta}^{\alpha}(z) g_{jk\mu}^{\lambda}(z) \zeta_k^{\beta\mu}.$$

(3)   *Dual* bundle F^* of F. This is the bundle defined by $\{f_{jk}^*\}$ where $f_{jk}^* = (f_{jk}^{-1})^t = (f_{kj})^t$ which is the transposed inverse of f_{jk}. Then $(z, \zeta_j^{*\alpha})$ is identified with $(z, \zeta_k^{*\beta})$ if and only if

$$\zeta_j^{*\alpha} = \sum f_{jk\beta}^{*\alpha}(z) \zeta_k^{*\beta} = \sum f_{kj\alpha}^{\beta}(z) \zeta_k^{*\beta}.$$

Sometimes we write $\zeta_{j\alpha}^*$ for $\zeta_j^{*\alpha}$. Then we have

$$\zeta_{j\alpha}^* = \sum_{\beta=1}^{n} f_{jk\alpha}^{\beta}(z) \zeta_{k\beta}^*.$$

(4)   *Complex Conjugate* $\bar{F}$ of F. This bundle is defined by the cocycle $\{\bar{f}_{jk}\}$.

Let us now define subbundles and quotient bundles. Suppose that, by a suitable choice of $\mathcal{U} = \{U_j\}$ and of fibre coordinates ζ_j^*, the matrices f_{jk} in the 1-cocycle $\{f_{jk}\}$ defining F can be written as follows:

$$f_{jk} = \begin{pmatrix} A_{jk} & B_{jk} \\ 0 & C_{jk} \end{pmatrix}.$$

Hence $f^{\alpha}_{jk\beta}(z) = 0$ for $1 \leq \beta \leq m$, $m + 1 \leq \alpha \leq n$. Thus $\zeta^{\alpha}_j = \sum^n_{\beta = m+1} f^{\alpha}_{jk\beta} \zeta^{\beta}_k$ for $\alpha > m$, and if $\zeta^{\beta}_k = 0$ for $\beta > m$, then $\zeta^{\alpha}_j = 0$ for $\alpha > m$. Let $F' = \cup\, U_j \times \mathbb{C}^m$ where $\mathbb{C}^m = \{(\zeta^1_j, \cdots, \zeta^m_j, 0, \cdots, 0)\} \subseteq \mathbb{C}^n$, and we identify (z, ζ_j) and (z, ζ_k) if $\zeta_j = A_{jk}(z)\zeta_k$. Then F' is a *subbundle* of F. The *quotient bundle* $F'' = F/F'$ is a $\mathbb{C}^{n-m}$ bundle defined by the 1-cocycle $\{C_{jk}\}$.

DEFINITION 5.5. A holomorphic (or differentiable) *section* of F over $U \subseteq M$ is a holomorphic (differentiable) map $\varphi : z \to \varphi(z)$ of $U \to F$ such that $\pi\varphi(z) = z$ where F is a holomorphic (or differentiable) $\mathbb{C}^n$ bundle.

We see that locally φ is a set of n-functions. Since local sections and germs of sections are defined, we get a sheaf of germs of sections of F. We denote by $\mathcal{O}(F)$ (or $\mathcal{D}(F)$) for sheaf over M of germs of holomorphic (or differentiable) sections of F. Then locally, $\mathcal{O}(F) = \mathcal{O}\,|\,U_j \oplus \cdots \oplus \mathcal{O}\,|\,U_j$ (sum n-times), where $\mathcal{O}\,|\,U_j$ means $\mathcal{O}$ restricted to U_j and $\mathcal{O}_z(F) = \mathcal{O}_z \oplus \cdots \oplus \mathcal{O}_z$ (n-times).

We now review tangent bundles and tensor bundles. Let M be a complex manifold and $\{U_j\}$ an open covering of M with coordinate patches with coordinates $(z^1_j, \cdots, z^n_j)$ on U_j. A (*holomorphic*) *tangent vector* at z is an element of the form $v = \sum^n_{\alpha = 1} \zeta^{\alpha}_j(\partial/\partial z^{\alpha}_j)$. It is easy to see that the set $T_z(M)$ of all complex tangent vectors at z is a complex vector space $T_z(M) \cong \mathbb{C}^n$. If $z \in U_k$ another chart at z, then we identify $\sum^n_{\beta = 1} \zeta^{\beta}_k(\partial/\partial z^{\beta}_k)$ with $\sum \zeta^{\alpha}_j(\partial/\partial z^{\alpha}_j)$ if

$$\zeta^{\alpha}_j = \sum^n_{\beta = 1} f^{\alpha}_{jk\beta}(z)\zeta^{\alpha}_k, \qquad f^{\alpha}_{jk\beta}(z) = \frac{\partial z^{\alpha}_j}{\partial z^{\beta}_k}.$$

This is a linear identification so the vector space structure of $T_z(M)$ is well defined. The set $T(M) = \bigcup_{z \in M} T_z(M)$ is a complex analytic vector bundle defined by the 1-cocycle $\{f^{\alpha}_{jk\beta}(z)\}$. $T(M)$ is the *holomorphic tangent bundle* of M. $\overline{T(M)}$ is the *conjugate* (holomorphic) *tangent bundle* of M. And $\mathcal{T}(M) = T(M) \oplus \overline{T(M)}$ is the (complexified) *tangent bundle* of M. Then Θ, the sheaf of germs of holomorphic vector fields, is $\mathcal{O}(T(M))$.

If M is a differentiable manifold with local coordinates $(x^1_j, \cdots, x^n_j)$, then the (complex) tangent bundle $\mathcal{T}(M) = \bigcup_{x \in M} \mathcal{T}_x(M)$, where

$$\mathcal{T}_x(M) = \left\{ v \,|\, v = \sum^n_{\alpha = 1} \zeta^{\alpha}_j\left(\frac{\partial}{\partial x^{\alpha}_j}\right),\ \zeta^{\alpha}_j \in \mathbb{C}\right\}.$$

The real tangent bundle $\mathcal{T}_{\mathbb{R}}(M) = \bigcup_{x \in M} \mathcal{T}_{x_{\mathbb{R}}}(M)$, where

$$\mathcal{T}_{x_{\mathbb{R}}}(M) = \left\{ v \,|\, v = \sum^n_{\alpha = 1} \xi^{\alpha}_j\left(\frac{\partial}{\partial x^{\alpha}_j}\right),\ \xi^{\alpha}_j \in \mathbb{R}\right\}.$$

The relation is $\mathcal{T}(M) = \mathcal{T}_{\mathbb{R}}(M) \otimes_{\mathbb{R}} \mathbb{C}$, where $\mathbb{C}$ is the trivial (complex) line bundle over M considered as a real bundle.

Let $T^*(M)$ be the dual bundle of $T(M)$. If $T_z^*(M) = \{(\zeta_{j1}^*, \cdots, \zeta_{jn}^*)\}$, then the transition relations are

$$\zeta_{j\alpha}^* = \sum_{\beta=1}^{n} \frac{\partial z_k^\beta}{\partial z_j^\alpha} \zeta_{k\beta}^*.$$

We use the following notation: An element $u \in T_z^*(M)$ shall be written $u = \sum_\alpha \zeta_{j\alpha}^* \, dz_j^\alpha$, where $dz_j^\alpha(\partial/\partial z_j^\beta) = \delta_\beta^\alpha$ as an element of $T_z^*(M)$. Briefly let $T = T(M)$, $T^* = T^*(M)$. A *tensor bundle* is a bundle of the form

$$T \otimes \cdots \otimes T \otimes T^* \otimes \cdots \otimes \overline{T} \otimes \cdots \otimes \overline{T}^*.$$

We denote $T \otimes \cdots \otimes T = (\otimes T)^p$ for the p-fold tensor product of T. We remark that $\overline{T}$ is *not* a holomorphic bundle so $(\otimes T)^p \otimes (\otimes T^*)^q$ is a holomorphic bundle but $(\otimes T)^p \otimes (\otimes T^*)^q \otimes (\otimes \overline{T})^r \otimes (\otimes \overline{T}^*)^s$ is only differentiable. A holomorphic (differentiable) tensor field is a holomorphic (differentiable) section of a tensor bundle.

We now give a brief treatment of differential forms.

DEFINITION 5.6.   A *differential form* of type (p, q) (or a (p, q)-form) over an open set $W \subseteq M$ is a differentiable section $\varphi : z \to [z, \varphi_{j\alpha_1 \cdots \alpha_p \bar\beta_1 \cdots \bar\beta_q}(z)]$ of $(\otimes T^*)^p \otimes (\overline{T}^*)^q$ over W such that the fibre coordinates $\varphi_{j\alpha_1 \cdots \alpha_p \bar\beta_1 \cdots \bar\beta_q}$ are skew-symmetric with respect to $\alpha_1 \cdots \alpha_p \bar\beta_1 \cdots \bar\beta_q$. If $p = 1$, $q = 0$, $\varphi(z) = \sum_{\alpha=1}^{n} \varphi_{j\alpha}(z) \, dz_j^\alpha$. In general, we represent the (p, q) form as follows:

$$\varphi(z) = \frac{1}{p!\,q!} \sum_{\substack{\alpha_1, \cdots, \alpha_p \\ \bar\beta_1, \cdots, \bar\beta_q}} \varphi_{j\alpha_1 \cdots \alpha_p \bar\beta_1 \cdots \bar\beta_q}(z) \, dz_j^{\alpha_1} \wedge \cdots \wedge dz_j^{\alpha_p} \wedge \cdots \wedge d\bar z_j^{\bar\beta_q},$$

where $\overline{dz^{\beta_i}} = d\bar z^{\beta_i}$ and "$\wedge$" is the wedge product of skew-symmetric forms and satisfies for example, $dz^\alpha \wedge dz^\beta = -dz^\beta \wedge dz^\alpha$. [*Note*: We write $\bar z^\alpha = \overline{z^\alpha} = z^{\bar\alpha}$.]

If M is only differentiable we still have $\mathcal{T}(M)$ and $\mathcal{T}^*(M)$. Then if W is open in M, we make the following definition:

DEFINITION 5.6′.   A differential form of degree p over W

$$\varphi = \frac{1}{p!} \sum \varphi_{j\alpha_1 \cdots \alpha_p}(x) \, dx_j^{\alpha_1} \wedge \cdots \wedge dx_j^{\alpha_p}$$

is a section over W of $\mathcal{T}^p$ which is skew-symmetric in the indices $\alpha_1 \cdots \alpha_p$. If φ is a p-form and x is a q-form, we define the *wedge product* of forms

$$\varphi \wedge \psi = \frac{1}{p!\,q!} \sum \varphi_{j\alpha_1 \cdots \alpha_p} \psi_{j\beta_1 \cdots \beta_q} \, dx_j^{\alpha_1} \wedge \cdots \wedge dx_j^{\alpha_p} \wedge dx_j^{\beta_1} \wedge \cdots \wedge dx_j^{\beta_q}.$$

For example, if $\varphi = \frac{1}{2} \sum \varphi_{\alpha\beta}\, dx^\alpha \wedge dx^\beta$ and $\psi = \sum \psi_\gamma\, dx^\gamma$, then

$$\varphi \wedge \psi = \frac{1}{2} \sum \varphi_{\alpha\beta} \psi_\gamma\, dx^\alpha \wedge dx^\beta \wedge dx^\gamma$$

$$= \frac{1}{3!} \sum \chi_{\alpha\beta\gamma}\, dx^\alpha \wedge dx^\beta \wedge dx^\gamma,$$

where

$$\chi_{\alpha\beta\gamma} = \varphi_{\alpha\beta} \psi_\gamma - \varphi_{\alpha\gamma} \psi_\beta + \varphi_{\beta\gamma} \psi_\alpha.$$

Hence

$$(\varphi \wedge \psi)_{\alpha\beta\gamma} = \varphi_{\alpha\beta} \psi_\gamma - \varphi_{\alpha\gamma} \psi_\beta + \varphi_{\beta\gamma} \psi_\alpha.$$

Definition 5.7. If

$$\varphi = \frac{1}{p!} \sum_{\alpha_1 \cdots \alpha_p} \varphi_{j\alpha_1 \cdots \alpha_p}\, dx_j^{\alpha_1} \wedge \cdots \wedge dx_j^{\alpha_p},$$

then the *exterior derivative* $d\varphi$ is

$$d\varphi = \frac{1}{p!} \sum_{\alpha, \alpha_1, \cdots, \alpha_p} \frac{\partial \varphi_{j\alpha_1 \cdots \alpha_p}}{\partial x_j^\alpha}\, dx_j^\alpha \wedge dx_j^{\alpha_1} \wedge \cdots \wedge dx_j^{\alpha_p}$$

$$= \frac{1}{(p+1)!} \sum_{\alpha_0, \cdots, \alpha_p} \psi_{j\alpha_0 \cdots \alpha_p}\, dx_j^{\alpha_0} \wedge \cdots \wedge dx_j^{\alpha_p}, \tag{1}$$

where

$$\psi_{\alpha_0 \cdots \alpha_p} = \left(\frac{\partial}{\partial x^{\alpha_0}}\right)\varphi_{\alpha_1 \cdots \alpha_p} - \left(\frac{\partial}{\partial x^{\alpha_1}}\right)\varphi_{\alpha_0\alpha_2 \cdots \alpha_p} + \cdots$$

$$+ (-1)^p \left(\frac{\partial}{\partial x^{\alpha_p}}\right)\varphi_{\alpha_0 \cdots \alpha_{p-1}}.$$

Proposition 5.3. $d\varphi$ is well defined (that is, the definition is independent of the choice of local coordinates x_j).

Proof. We write out the proof for $p = 2$ and leave the general case to the reader. So consider

$$\varphi = \frac{1}{2} \sum \varphi_{j\alpha\beta}\, dx_j^\alpha \wedge dx_j^\beta = \frac{1}{2} \sum \varphi_{k\lambda\mu}\, dx_k^\lambda \wedge dx_k^\mu$$

on $U_j \cap U_k$. We want to see that

$$\frac{1}{2} \sum \frac{\partial \varphi_{j\alpha\beta}}{\partial x_j^\gamma}\, dx_j^\gamma \wedge dx_j^\alpha \wedge dx_j^\beta = \frac{1}{2} \sum \frac{\partial \varphi_{k\lambda\mu}}{\partial x_k^\nu}\, dx_k^\nu \wedge dx_k^\lambda \wedge dx_k^\mu. \tag{2}$$

The following rules of transformation are given to us.

$$\varphi_{j\alpha\beta} = \sum_{\lambda,\mu} \frac{\partial x_k^\lambda}{\partial x_j^\alpha} \frac{\partial x_k^\mu}{\partial x_j^\beta} \varphi_{k\lambda\mu},$$

$$dx_k^\lambda = \sum_\alpha \frac{\partial x_k^\lambda}{\partial x_j^\alpha} dx_j^\alpha.$$

Thus on $U_j \cap U_k$ the left-hand side $\natural$ of Equation (2) is

$$\natural = \tfrac{1}{2} \sum_{\alpha,\beta,\gamma} \left(\frac{\partial^2 x_k^\lambda}{\partial x_j^\alpha \partial x_j^\gamma} \cdot \frac{\partial x_k^\mu}{\partial x_j^\beta} \varphi_{k\lambda\mu} + \frac{\partial x_k^\lambda}{\partial x_j^\alpha} \frac{\partial^2 x_k^\mu}{\partial^2 x_j^\gamma \partial x_j^\beta} \varphi_{k\lambda\mu} \right.$$

$$\left. + \frac{\partial x_k^\lambda}{\partial x_j^\alpha} \frac{\partial x_k^\mu}{\partial x_j^\beta} \frac{\partial \varphi_{k\lambda\mu}}{\partial x_j^\gamma} \right) dx_j^\gamma \wedge dx_j^\alpha \wedge dx_j^\beta.$$

The sum of the first two terms is zero, so

$$\natural = \tfrac{1}{2} \sum \frac{\partial \varphi_{k\lambda\mu}}{\partial x_k^\mu} \frac{\partial x_k^\mu}{\partial x_j^\gamma} dx_j^\gamma \wedge \frac{\partial x_k^\lambda}{\partial x_j^\alpha} dx_j^\alpha \wedge \frac{\partial x_k^\mu}{\partial x_j^\beta} dx_j^\beta$$

$$= \tfrac{1}{2} \sum \frac{\partial \varphi_{k\lambda\mu}}{\partial x_k^\mu} dx_k^\nu \wedge dx_k^\lambda \wedge dx_k^\mu,$$

and (2) is proved.   Q.E.D.

REMARK.   For other types of tensor fields such a "nice" operator does not exist. To correct for the difficulty one introduces the idea of a connection.

PROPOSITION 5.4. $dd\varphi = 0$.

Proof. $d\varphi = \dfrac{1}{p!} \sum \dfrac{\partial \varphi_{\alpha_1 \cdots \alpha_p}}{\partial x^\alpha} dx^\alpha \wedge \cdots \wedge dx^{\alpha_p}.$

$$dd\varphi = \frac{1}{p!} \sum \frac{\partial^2 \varphi_{\alpha_1 \cdots \alpha_p}}{\partial x^\beta \partial x^\alpha} dx^\beta \wedge dx^\alpha \wedge \cdots \wedge dx^{\alpha_p}$$

$$= 0,$$

since $\partial^2 \varphi_{\alpha_1 \cdots \alpha_p}/\partial x^\beta \partial x^\alpha$ is symmetric in α, β and $dx^\beta \wedge dx^\alpha$ is skew-symmetric.

PROPOSITION 5.5.   If φ is a p-form, then

$$d(\varphi \wedge \psi) = d\varphi \wedge \psi + (-1)^p \varphi \wedge d\psi.$$

Proof. Easy.

THEOREM 5.1. (Poincaré's Lemma)   Suppose that a p-form φ, $p \geq 1$, satisfies $d\varphi = 0$ on a neighborhood U of $0 = (0, \cdots, 0) \subseteq \mathbb{R}^n$. Then there is a $(p-1)$-form ψ on a neighborhood W, $0 \in W \subseteq U$, such that $\varphi = d\psi$ (locally $d\varphi = 0$ if and only if $\varphi = d\psi$ for some ψ).

Proof.   Let $\varphi = \varphi(x) = \dfrac{1}{p!} \sum \varphi_{\alpha_1 \cdots \alpha_p}(x)\, dx^{\alpha_1} \wedge \cdots \wedge dx^{\alpha_p}$.

We fix p and prove the theorem by induction on the dimension n. The first step is the case $n = p$, and $\varphi = \varphi_{12 \cdots p}(x)\, dx^1 \wedge \cdots \wedge dx^p$. Define

$$\psi(x) = g(x)\, dx^1 \wedge \cdots \wedge dx^{p-1}$$

by

$$g(x^1, \cdots, x^p) = (-1)^{p-1} \int_0^{x^p} \varphi_{1 \cdots p}(x^1, \cdots, x^{p-1}, t)\, dt,$$

where W is a star-shaped neighborhood of 0 and $(x^1, \cdots, x^p) \in W$ (that is, if $x \in W$ so is the line joining 0 and x). Then

$$d\psi = \sum \frac{\partial g(x)}{\partial x^\alpha}\, dx^\alpha \wedge dx^1 \wedge \cdots \wedge dx^{p-1}$$

$$= \frac{\partial g(x)}{\partial x^p}\, dx^p \wedge dx^1 \wedge \cdots \wedge dx^{p-1}$$

$$= \varphi_{1 \cdots p}(x^1, \cdots, x^p)\, dx^1 \wedge \cdots \wedge dx^p$$

$$= \varphi.$$

Now assume the result for $n = m - 1$, $m > p$ and consider $n = m$. Then

$$\varphi(x) = \frac{1}{p!} \sum_{1 \leq \alpha_\lambda \leq m-1} \varphi_{\alpha_1 \cdots \alpha_p}\, dx^{\alpha_1} \wedge \cdots \wedge dx^{\alpha_p}$$

$$+ \frac{1}{(p-1)!} \sum_{1 \leq \alpha_\lambda \leq m-1} \varphi_{m\alpha_1 \cdots \alpha_{p-1}}\, dx^m \wedge \cdots \wedge dx^{\alpha_{p-1}}.$$

Let

$$g_{\alpha_1 \cdots \alpha_{p-1}}(x^1, \cdots, x^m) = \int_0^{x^m} \varphi_{m\alpha_1 \cdots \alpha_{p-1}}(x^1, \cdots, x^{m-1}, t)\, dt$$

and

$$\psi = \frac{1}{(p-1)!} \sum_{1 \leq \alpha_\lambda \leq m-1} g_{\alpha_1 \cdots \alpha_{p-1}}\, dx^{\alpha_1} \wedge \cdots \wedge dx^{\alpha_{p-1}}.$$

Then

$$d\psi = \frac{1}{(p-1)!} \sum_{1 \le \alpha_\lambda \le m-1} \sum_{\alpha=1}^{m-1} \frac{\partial g}{\partial x^\alpha} dx^\alpha \wedge dx^{\alpha_1} \wedge \cdots \wedge dx^{\alpha_{p-1}}$$

$$+ \frac{1}{(p-1)!} \sum_{1 \le \alpha_\lambda \le m-1} \varphi_{m\alpha_1 \cdots \alpha_{p-1}} dx^m \wedge dx^{\alpha_1} \wedge \cdots \wedge dx^{\alpha_{p-1}}.$$

Hence

$$\varphi(x) - d\psi(x) = \frac{1}{p!} \sum_{\alpha_\lambda=1}^{m-1} \chi_{\alpha_1 \cdots \alpha_p}(x) \, dx^{\alpha_1} \wedge \cdots \wedge dx^{\alpha_p}$$

$$= \chi$$

and

$$d\chi = d\varphi - dd\psi = d\varphi = 0.$$

But

$$d\chi = \frac{1}{p!} \sum^{m-1} \frac{\partial \chi_{\alpha_1 \cdots \alpha_p}}{\partial x^m} dx^m \wedge dx^{\alpha_1} \wedge \cdots \wedge dx^{\alpha_p}$$

$$+ \text{ terms not involving } dx^m.$$

So $\partial\chi_{\alpha_1 \cdots \alpha_p}/\partial x^m = 0$ and this implies that χ is independent of x^m, $\chi = \chi(x^1, \cdots, x^{m-1})$. By induction $\chi = d\sigma$ and thus $\varphi = d(\psi + \sigma)$. Q.E.D.

We denote the sheaf over M of germs of p-forms (C^∞ p-forms) by A^p. Then we have:

Corollary. (to Poincaré's Lemma) Let $\varphi \in A^p$, $p \ge 1$. Then $d\varphi = 0$, if and only if $\varphi = d\psi$, $\psi \in A^{p-1}$.

Proof. We need only remark that d can be defined on the germ level and then use Theorem 5.1.

We notice that $A^0 = \mathscr{D}$, the sheaf of germs of C^∞ differentiable functions; and $\varphi \in A^0$, $d\varphi = 0$ if and only if φ is locally constant. Thus:

Theorem 5.2. The following sequence (of sheaves over M) is exact

$$0 \longrightarrow C \longrightarrow A^0 \overset{d}{\longrightarrow} A^1 \overset{d}{\longrightarrow} \cdots \overset{d}{\longrightarrow} A^n \overset{d}{\longrightarrow} 0,$$

where $n = \dim M$.

Theorem 5.3. The sheaf A^p is a fine sheaf.

Proof. Given any locally finite covering $\{U_j\}$ of M, we have a partition of unity $\{\rho_j\}$,

$$\sum \rho_j = 1 \quad \text{and} \quad \overline{\{x \mid \rho_j(x) \neq 0\}} \subseteq U_j,$$

where $\rho_j(x)$ is a C^∞ function on M with $\rho_j(x) \geq 0$. Define a homomorphism $h_j : A^p \to A^p$ by

$$h_j(\varphi) = \rho_j(x) \cdot \varphi.$$

Then $\sum h_j = id$ and $h_j(A_x^p) = 0$ if $\rho_j(x) = 0$.   Q.E.D.

THEOREM 5.4.   (de Rham's Theorem)   Let dA^p be the image of A^p under d. Then $d\Gamma(M, A^{q-1}) \subseteq \Gamma(M, dA^{q-1})$ and

$$H^q(M, \mathbb{C}) \cong \frac{H^0(M, dA^{q-1})}{dH^0(M, A^{q-1})}.$$

Proof. We note that $dA^{p-1} \subset A^p$ is the subsheaf of A^p which consists of germs of p-forms φ such that $d\varphi = 0$. A form φ is called *d-closed* if $d\varphi = 0$. A form φ is *exact* if $\varphi = d\psi$. Thus Theorem 5.5 says that the closed p-forms modulo the exact p-forms is isomorphic to the cohomology with complex coefficients.   For the proof we use Theorems 5.4, 5.3, 4.3, and 4.1. Since

$$0 \to dA^{p-1} \to A^p \to dA^p \to 0$$

is exact for $p \geq 1$,

$$0 \to H^0(dA^{p-1}) \to H^0(A^p) \to H^0(dA^p) \to H^1(dA^{p-1}) \to \cdots$$

is exact. But $H^q(A^p) = 0$ for $q \geq 1$. Thus,

$$0 \to H^0(dA^{p-1}) \to H^0(A^p) \to H^0(dA^p) \to H^1(dA^{p-1}) \to 0 \tag{3}$$

is exact, and $H^{q-1}(dA^p) \cong H^q(dA^{p-1})$ for $q \geq 2$. Equation (3) gives

$$H^1(dA^{p-1}) \cong H^0(dA^p)/dH^0(A^p).$$

For $p = 0$ we have

$$0 \longrightarrow \mathbb{C} \longrightarrow A^0 \xrightarrow{\ d\ } dA^0 \longrightarrow 0.$$

Thus

$$0 \to H^0(\mathbb{C}) \to H^0(A^0) \to H^0(dA^0) \to H^1(\mathbb{C}) \to H^1(A^0)$$
$$\to H^1(dA^0) \to \cdots \to H^{q-1}(dA^0) \to H^q(\mathbb{C}) \to 0.$$

Since

$$H^q(A^0) = 0 \text{ for } q \geq 1, \ H^{q-1}(dA^0) \cong H^q(\mathbb{C}), \text{ and } H^1(\mathbb{C}) \cong H^0(dA^0)/dH^0(A^0)$$

for $q \geq 2$. Thus,

$$H^q(\mathbb{C}) \cong H^{q-1}(dA^0) \cong H^{q-2}(dA^1) \cong \cdots H^1(dA^{q-2})$$

$$\cong \frac{H^0(dA^{q-1})}{dH^0(A^{q-1})} \qquad \text{for } q \geq 2.$$

These two statements prove the theorem.   Q.E.D.

REMARK.   We have actually proved more. Let $\mathscr{S}$ be a sheaf over X. By a *fine resolution* of $\mathscr{S}$ we mean an exact sequence $(\mathscr{R})$,

$$(\mathscr{R})\; 0 \longrightarrow \mathscr{S} \xrightarrow{\;h\;} \mathscr{A}^0 \xrightarrow{\;h\;} \mathscr{A}^1 \xrightarrow{\;h\;} \mathscr{A}^2 \longrightarrow \cdots$$

such that each $\mathscr{A}^p$ is fine.

*THEOREM 5.4.   If $(\mathscr{R})$ is a fine resolution of $\mathscr{S}$, then

$$H^q(X, \mathscr{S}) \cong \frac{H^0(X, h\mathscr{A}^{q-1})}{hH^0(X, \mathscr{A}^{q-1})} \qquad \text{for } q \geq 1.$$

Proof.   Same method.

6.   A Theorem of Dolbeault (A fine resolution of $\mathcal{O}$)

Let M be a complex manifold with covering $\{U_j\}$ by coordinate patches where $(z_j^1, \cdots, z_j^n)$ are local complex coordinates on U_j. Let

$$\varphi = \varphi(z) = \frac{1}{p!\,q!} \sum \varphi_{\alpha_1 \cdots \bar{\beta}_1 \cdots \bar{\beta}_q}(z)\, dz^{\alpha_1} \wedge \cdots \wedge dz^{\bar{\beta}_q}.$$

DEFINITION 6.1.

$$\partial\varphi = \frac{1}{p!\,q!} \sum_{\alpha, \alpha_1, \cdots \bar{\beta}_q} \frac{\partial \varphi_{\alpha_1 \cdots \bar{\beta}_q}}{\partial z^{\alpha}}\, dz^{\alpha} \wedge dz^{\alpha_1} \wedge \cdots \wedge dz^{\bar{\beta}_q},$$

where

$$\frac{\partial}{\partial z^{\alpha}} = \frac{1}{2}\left(\frac{\partial}{\partial x^{\alpha}} - i\,\frac{\partial}{\partial y^{\alpha}} \right),$$

and

$$\bar{\partial}\varphi = \frac{1}{p!\,q!} \sum_{\alpha_1 \cdots \bar{\beta}, \bar{\beta}_1 \cdots \bar{\beta}_q} \frac{\partial}{\partial z^{\bar{\beta}}}\, \varphi_{\alpha_1 \cdots \bar{\beta}_q}\, dz^{\bar{\beta}} \wedge dz^{\alpha} \wedge \cdots \wedge dz^{\bar{\beta}_q},$$

where

$$\frac{\partial}{\partial z^{\beta}} = \frac{1}{2}\left(\frac{\partial}{\partial x^{\beta}} + i\,\frac{\partial}{\partial y^{\beta}}\right).$$

REMARK.   It is easy to verify that

$$\partial\partial = 0, \qquad \bar{\partial}\bar{\partial} = 0. \tag{1}$$

From the complex coordinates we get real coordinates x^{γ} defined by $z^{\alpha} = x^{2\alpha-1} + i\,x^{2\alpha}$. Then

$$dz^{\alpha} = dx^{2\alpha-1} + i\,dx^{2\alpha}$$

$$dz^{\bar{\alpha}} = dx^{2\alpha-1} - i\,dx^{2\alpha}.$$

So

$$\varphi = \frac{1}{(p+q)!}\sum_{\lambda}\varphi^{*}_{\lambda_1,\cdots,\lambda_{p+q}}\,dx^{\lambda_1}\wedge\cdots\wedge dx^{\lambda_{p+q}}.$$

LEMMA 6.1. $d = \partial + \bar{\partial}$.

Proof.   We easily check that

$$dz\,\frac{\partial}{\partial z} + d\bar{z}\,\frac{\partial}{\partial \bar{z}} = dx\,\frac{\partial}{\partial x} + dy\,\frac{\partial}{\partial y}.$$

In fact

$$\partial + \bar{\partial} = \sum_{\alpha=1}^{n}\left(dz^{\alpha}\,\frac{\partial}{\partial z^{\alpha}} + dz^{\bar{\alpha}}\,\frac{\partial}{\partial z^{\bar{\alpha}}}\right) = \sum_{\lambda=1}^{2n}dx^{\lambda}\,\frac{\partial}{\partial x^{\lambda}} = d. \qquad \text{Q.E.D.}$$

COROLLARY. $\partial\,\bar{\partial} = -\bar{\partial}\,\partial$.

Proof. $0 = dd = (\partial + \bar{\partial})(\partial + \bar{\partial}) = \bar{\partial}\,\partial + \partial\,\bar{\partial}$.

THEOREM 6.1.   [Dolbeault's lemma (an analogue of Poincaré's lemma)] If *a* C^{∞} $(0, q)$-form φ satisfies $\bar{\partial}\varphi = 0$ on a neighborhood $U \subseteq \mathbb{C}^n$ of z_0, then there is a C^{∞} $(0, q-1)$-form ψ on W with $z_0 \in W \subseteq U$ (W open) such that $\varphi = \bar{\partial}\psi$ on W.

First, we turn to:

LEMMA 6.2.   Let $f(z)$ be a bounded C^{∞} function on $U \subseteq \mathbb{C}$.   Suppose

$$g(z) = \frac{-1}{\pi}\iint\limits_{U}\frac{f(\zeta)\,d\xi\,d\eta}{\zeta - z},$$

where $\zeta = \xi + i\eta$. Then g is C^{∞} on U and $\partial g/\partial \bar{z} = f$.

Proof.   Pick $z_0 \in U$ and let

$$\Delta_1 = \{z|\ |z - z_0| < \varepsilon\}, \qquad \Delta_2 = \{z|\ |z - z_0| < 2\varepsilon\}.$$

Suppose $z \in \Delta_1$. Let $f_1(\zeta)$ be a C^∞ function which is identically equal to f in Δ_1 and is zero outside of Δ_2. Then

$$f(\zeta) = f_1(\zeta) + f_2(\zeta),$$

where

$$f_1(\zeta) = 0 \text{ for } \zeta \notin \Delta_2,$$
$$f_2(\zeta) = 0 \text{ for } \zeta \in \Delta_1.$$

Then we have $g(z) = g_1(z) + g_2(z)$, where

$$g_\nu(z) = \frac{-1}{\pi} \iint_U \frac{f_\nu(z)\, d\xi\, d\eta}{\zeta - z}.$$

For $z \in \Delta_1$, $g_2(z)$ is holomorphic, so $(\partial/\partial\bar{z})\, g_2(z) = 0$. Since $f_1 = 0$ outside of Δ_1 we can consider

$$g_1(z) = \frac{-1}{\pi} \iint_{\mathbb{C}} \frac{f_1(\zeta)\, d\xi\, d\eta}{\zeta - z}.$$

Let

$$\zeta - z = w = s + i\, t = re^{i\theta}.$$

Then

$$g_1(z) = -\frac{1}{\pi} \iint_{\mathbb{C}} \frac{f_1(w + z)}{w}\, ds\, dt = -\frac{1}{\pi} \iint_{\mathbb{C}} \frac{f_1(re^{i\theta} + z)r\, dr\, d\theta}{re^{i\theta}}$$

$$= -\frac{1}{\pi} \iint_{\mathbb{C}} f_1(re^{i\theta} + z)e^{-i\theta}\, dr\, d0.$$

Thus $g_1(z)$ is C^∞ in z. Then

$$\frac{\partial}{\partial\bar{z}} g_1 = \frac{-1}{\pi} \iint \frac{\partial}{\partial\bar{z}} f_1(re^{i\theta} + z)e^{-i\theta}\, dr\, d\theta$$

$$= \frac{-1}{\pi} \iint \frac{\partial}{\partial\bar{z}} f_1(w + z) \frac{ds\, dt}{w}$$

and

$$\frac{\partial}{\partial\bar{z}} f_1(w + z) = \frac{\partial}{\partial\bar{w}} f_1(w + z).$$

We use Stokes' theorem. If W is a domain in $\mathbb{C}$ with boundary C a Jordan curve, then

$$\int_C u\, dt - v\, ds = \iint_W \left(\frac{\partial u}{\partial s} + \frac{\partial u}{\partial t} \right) ds\, dt.$$

If $\varphi = u\,dt - v\,ds$, then $d\varphi = \left(\dfrac{\partial u}{\partial s} + \dfrac{\partial u}{\partial t} \right) ds \wedge dt$ so

$$\int_C \varphi = \int_W d\varphi.$$

We notice that

$$dw \wedge d\overline{w} = (ds + idt) \wedge (ds - idt) = -2i\,ds \wedge dt.$$

Hence,

$$\frac{\partial}{\partial \overline{z}} g_1(z) = \frac{1}{2\pi i} \iint \frac{\partial}{\partial \overline{w}} f_1(w + z) \frac{dw \wedge d\overline{w}}{w} = \frac{-1}{2\pi i} \iint_C d\left[f_1(w + z) \frac{dw}{w} \right],$$

since

$$d\left(f_1 \frac{dw}{w} \right) = (\partial + \overline{\partial})\left(f_1 \frac{dw}{w} \right)$$

$$= \frac{\partial f_1}{\partial w} dw \wedge \frac{dw}{w} + \frac{\partial f_1}{\partial \overline{w}} d\overline{w} \wedge \frac{dw}{w}$$

$$= -\frac{\partial f_1}{\partial \overline{w}} \frac{dw \wedge d\overline{w}}{w}.$$

Thus, if we set

$$\Gamma = \{z \mid |z - z_0| = 2\varepsilon\},$$
$$\gamma = \{z \mid |z - z_0| = \varepsilon\},$$

then

$$\frac{\partial}{\partial \overline{z}} g_1(z) = -\frac{1}{2\pi i} \left(\int_\Gamma f_1 \frac{dw}{w} - \int_\gamma f_1 \frac{dw}{w} \right) - \frac{1}{2\pi i} \iint_{\Delta_1} d\left(f_1 \frac{dw}{w} \right)$$

$$= \frac{1}{2\pi i} \int_\gamma f_1(w + z) \frac{dw}{w} - \frac{1}{2\pi i} \iint_{\Delta_1} d\left(f_1 \frac{dw}{w} \right).$$

Taking the limit as $\varepsilon \to 0$ we see that

$$\frac{\partial g}{\partial \overline{z}} = \frac{\partial g_1}{\partial \overline{z}} = f_1(z) = f(z). \qquad \text{Q.E.D.}$$

Instead of assuming that f is bounded we could have proved, using the same proof:

LEMMA 6.2.   If $f(z)$ is C^∞ for $|z| < R$, then

$$g(z) = -\frac{1}{\pi} \iint\limits_{\xi^2 + \eta^2 < R - \varepsilon} \frac{f(\zeta)}{\zeta - z}\, d\xi\, d\eta$$

is C^∞ for $|z| < R - \varepsilon$ and

$$\frac{\partial g(z)}{\partial \bar{z}} = f(z) \qquad \text{for } |z| < R - \varepsilon,$$

where $\varepsilon > 0$ and $R - \varepsilon > 0$.

Proof.   (of Theorem 6.1)   We may as well assume

$$U = U_R = \{z|\; |z| < R, \text{ where } |a| = \max|z^\alpha|\}.$$

Then we want to prove that if $\bar{\partial}\varphi = 0$,

$$\varphi = \frac{1}{q!} \sum \varphi_{\beta_1 \cdots \beta_q}\, d\bar{z}^{\beta_1} \wedge \cdots \wedge d\bar{z}^{\beta_q}$$

is a $C^\infty (0, q)$-form on U_R, then for any $\varepsilon > 0$, $R - \varepsilon > 0$, there is a $C^\infty (q - 1)$-form ψ on $U_{R-\varepsilon}$ such that

$$\bar{\partial}\psi = \varphi \text{ on } U_{R-\varepsilon} \qquad (q \geq 1).$$

Suppose

$$\varphi = \frac{1}{q!} \sum_{\beta_i \leq m} \varphi_{\beta_1 \cdots \beta_q}(z)\, d\bar{z}^{\beta_1} \wedge \cdots \wedge d\bar{z}^{\beta_q}.$$

By this we mean that form φ does not involve differentials of coordinates z^i for $i > m$. The proof will be by induction on m with fixed q, n. First we consider $m = q$. Then

$$\varphi = \varphi_{12 \cdots q}\, d\bar{z}^1 \wedge \cdots \wedge d\bar{z}^q$$

and

$$0 = \bar{\partial}\varphi = \sum_{\alpha=1}^{n} \frac{\partial}{\partial \bar{z}^\alpha} \varphi_{1 \cdots q}(z)\, d\bar{z}^\alpha \wedge d\bar{z}^1 \wedge \cdots \wedge d\bar{z}^q$$

$$= \sum_{\alpha \geq q+1} \frac{\partial}{\partial \bar{z}^\alpha} \varphi_{1 \cdots q}(z)\, d\bar{z}^\alpha \wedge d\bar{z}^1 \wedge \cdots \wedge d\bar{z}^q.$$

Hence, $(\partial/\partial \bar{z}^\alpha)\varphi_{1 \cdots q} = 0$ for $\alpha \geq q + 1$. Define

$$g(z^1, \cdots, z^n) = \frac{-1}{\pi} \iint\limits_{\xi^2 + \eta^2 < R - (\varepsilon/n)} \frac{\varphi_{1 \cdots q}(\zeta, z^2, \cdots, z^n)}{\zeta - z^1}\, d\xi\, d\eta.$$

Then g is C^∞ on $U_{R-(\varepsilon/n)}$ and $\partial g/\partial \bar{z}^1 = \varphi_{1\ldots q}(z)$. So let $\psi(z) = g(z)\, d\bar{z}^2 \wedge \cdots \wedge d\bar{z}^q$. Since $\partial\varphi/\partial\bar{z}^\alpha = 0$ for $\alpha \geq q + 1$, $\partial g/\partial\bar{z}^\alpha = 0$ for $\alpha \geq q + 1$. Thus,

$$\bar{\partial}\psi(z) = \sum_{\alpha=1}^{q} \frac{\partial g}{\partial\bar{z}^\alpha}\, d\bar{z}^\alpha \wedge d\bar{z}^2 \wedge \cdots \wedge d\bar{z}^q$$

$$= \frac{\partial g}{\partial\bar{z}^1}\, d\bar{z}^1 \wedge \cdots \wedge d\bar{z}^q$$

$$= \varphi.$$

Now assume the lemma is proved for $m \leq k - 1$ and consider $m = k$. So,

$$\varphi = \frac{1}{p!} \sum_{\beta_i \leq k} \varphi_{\beta_1\cdots\beta_q}\, d\bar{z}^{\beta_1} \wedge \cdots \wedge d\bar{z}^{\beta_q}$$

$$= \frac{1}{p!} \sum_{\beta_i \leq k-1} \varphi_{\beta_1\cdots\beta_q}\, d\bar{z}^{\beta_1} \wedge \cdots \wedge d\bar{z}^{\beta_q}$$

$$+ \frac{1}{(p-1)!} \sum_{\beta_i \leq k-1} d\bar{z}^k \wedge d\bar{z}^{\beta_2} \wedge \cdots \wedge d\bar{z}^{\beta_q}$$

and

$$0 = \bar{\partial}\varphi = \frac{1}{p!} \sum_{\beta_i \leq k} \sum_{\alpha=1}^{n} \frac{\partial}{\partial\bar{z}^\alpha} \varphi_{\beta_1\cdots\beta_q}\, d\bar{z}^\alpha \wedge \cdots \wedge d\bar{z}^{\beta_q}.$$

Thus, $0 = (\partial/\partial\bar{z}^\alpha)\, \varphi_{\beta_1\ldots\beta_q}(z) = 0$ for $\alpha \geq k + 1$. Define

$$g_{\beta_2\cdots\beta_q}(z) = \frac{-1}{\pi} \iint_{\xi^2\eta^2 < R-\epsilon_1} \varphi_{k\beta_2\cdots\beta_q}(z^1, \cdots, z^{k-1}, \zeta, z^{k+1}, \cdots, z^n)\, d\xi\, d\eta,$$

for $1 \leq \beta_2, \cdots, \beta_q \leq k - 1$. Then $g_{\beta_2\cdots\beta_q}$ is C^∞ on $U_{R-\varepsilon_1}$,

$$\frac{\partial}{\partial\bar{z}^k} g = \varphi_{k\beta_2\cdots\beta_q}, \qquad \text{and} \quad \frac{\partial g}{\partial\bar{z}^\alpha} = 0 \qquad \text{for } \alpha \geq k + 1.$$

Let

$$\omega = \frac{1}{(q-1)!} \sum_{\beta_i \leq k-1} g_{\beta_2\cdots\beta_q}\, d\bar{z}^{\beta_2} \wedge \cdots \wedge d\bar{z}^{\beta_q}.$$

Then

$$\bar{\partial}\omega = \frac{1}{(q-1)!} \sum_{\beta_i \leq k-1} \sum_{\alpha=1}^{k-1} \frac{\partial g}{\partial\bar{z}^\alpha}\, d\bar{z}^\alpha \wedge d\bar{z}^{\beta_2} \wedge \cdots \wedge d\bar{z}^{\beta_q}$$

$$+ \frac{1}{(q-1)!} \sum_{\beta_i \leq k-1} \varphi_{k\beta_2\cdots\beta_q}\, d\bar{z}^k \wedge d\bar{z}^{\beta_2} \wedge \cdots \wedge d\bar{z}^{\beta_q}.$$

Then

$$\sigma = \varphi - \bar{\partial}\omega = \frac{1}{p!} \sum_{\beta_i \leq k-1} \sigma_{\beta_1 \cdots \beta_q}\, d\bar{z}^{\beta_1} \wedge \cdots \wedge d\bar{z}^{\beta_q}$$

is C^∞ on $U_{R-\varepsilon_1}$. By induction there is a ψ which is C^∞ on $U_{R-\varepsilon_1-\varepsilon_2}$ such that $\sigma = \bar{\partial}\psi$. So $\varphi = \bar{\partial}(\omega + \psi)$ on $U_{R-\varepsilon_1-\varepsilon_2}$.   Q.E.D.

We can make an improvement.

THEOREM 6.2.   (also called Dolbeault's Lemma)   If φ is C^∞ on U_R and $\bar{\partial}\varphi = 0$, then there is a C^∞ ψ on U_R such that $\varphi = \bar{\partial}\psi$ on U_R.

Proof.   We first consider the case of a q-form φ with $q \geq 2$. Let $U_\nu = U_R - r/\nu$, $0 < r < R$. On each $U_{\nu+1}$ there is a ψ_ν such that $\delta\psi_\nu = \varphi$. We construct $\tilde{\psi}_1, \tilde{\psi}_2, \cdots$ by induction on ν such that $\bar{\partial}\tilde{\psi}_\nu = \varphi$ on $U_{\nu+1}$ and $\tilde{\psi}_\nu = \tilde{\psi}_{\nu+1}$ on U_ν. First set $\tilde{\psi}_1 = \psi_1$ on $U_2 = U_{R-(r/2)}$. Then $\bar{\partial}(\psi_2 - \psi_1) = 0$ on U_2. Hence there is a θ on $U_{R-(r/2)-\varepsilon}$ such that $\bar{\partial}\theta = \psi_2 - \psi_1$. Take $\rho(z) \geq 0$, C^∞ and such that

$$\rho(z) = \begin{cases} 1, & \text{on } U_1 \\ 0, & \text{outside } U_{R-(r/2)-2\varepsilon}. \end{cases}$$

Thus ρ, θ are defined on U_R, and

$$\rho\theta = \begin{cases} 0, & \text{outside } U_{R-(r/2)-2\varepsilon} \\ 0, & \text{on } U_1 = U_{R-r}. \end{cases}$$

Define $\tilde{\psi}_2 = \psi_2 - \bar{\partial}(\rho \cdot \theta)$. Then

$$\bar{\partial}\tilde{\psi}_2 = \bar{\partial}\psi_2 \text{ on } U_3$$

and

$$\tilde{\psi}_2 = \psi_2 - \bar{\partial}\theta = \psi_1 = \tilde{\psi}_1 \text{ on } U_1.$$

Suppose we have $\tilde{\psi}_1, \cdots, \tilde{\psi}_\nu$ such that $\bar{\partial}\tilde{\psi}_\lambda = \varphi$ on $U_{\lambda+1}$ for $\lambda \leq \nu$ and $\tilde{\psi}_\lambda = \tilde{\psi}_{\lambda+1}$ on U_λ for $\lambda \leq \nu - 1$. Then $\bar{\partial}(\psi_{\nu+1} - \tilde{\psi}_\nu) = 0$ on $U_{\nu+1} = U_{R-(r/\nu+1)}$ and $\psi_{\nu+1} - \tilde{\psi}_\nu = \bar{\partial}\theta$ for some θ on $U_{R-(r/\nu+1)-\varepsilon}$. Define $\tilde{\psi}_{\nu+1} = \psi_{\nu+1} - \bar{\partial}(\rho\theta)$ where

$$\rho = \begin{cases} 1, & \text{on } U_\nu \\ 0, & \text{outside } U_{R-(r/\nu+1)-2\varepsilon}. \end{cases}$$

Then $\bar{\partial}\tilde{\psi}_{\nu+1} = \varphi$ on $U_{\nu+2}$ and $\tilde{\psi}_{\nu+1} = \tilde{\psi}_\nu$ on U_ν. Then ψ is well defined, C^∞ on U_R, and $\bar{\partial}\psi = \bar{\partial}\tilde{\psi}_\nu = \varphi$.

Next consider the case $q = 1$. We assume ψ_ν on $U_{\nu+1}$ constructed so that $\bar{\partial}(\psi_{\nu+1} - \psi_\nu) = 0$ on $U_{\nu+1}$, $\varphi = \bar{\partial}\psi_\nu$ on $U_{\nu+1}$, and ψ_ν is a C^∞ function where φ is a given $(0,1)$-form such that $\bar{\partial}\varphi = 0$. Then $\bar{\partial}(\psi_{\nu+1} - \psi_\nu) = 0$ implies that

$\psi_{\nu+1} - \psi_\nu = f_\nu$ is a holomorphic function on $U_{\nu+1}$. Let $\tilde{\psi}_1 = \psi_1$. Since $\psi_2 - \tilde{\psi}_1 = f_1$ is holomorphic on U_2 let p_1 be a polynomial such that $|p_1 - f_1| < \frac{1}{2}$ on U_1. Set $\tilde{\psi}_2 = \psi_2 - p_1$ on U_3. In general, given $\tilde{\psi}_\nu$ such that $\bar{\partial}\psi_\nu = \bar{\partial}\tilde{\psi}_\nu$ on $U_{\nu+1}$ and $\bar{\partial}\psi_{\nu+1} = \varphi\,\bar{\partial}\tilde{\psi}_\nu$ on $U_{\nu+1}$, we can find a polynomial p_ν such that $|p_\nu - f_\nu| < 1/2\nu$ on U_ν since $\psi_{\nu+1} - \tilde{\psi}_\nu$ is holomorphic on $U_{\nu+1}$. Then $|\tilde{\psi}_{\nu+1} - \tilde{\psi}_\nu| < 1/2\nu$ on U_ν so the limit exists,

$$\lim_{\nu \to \infty} \tilde{\psi}_\nu = \psi$$

and

$$\psi = \psi_\nu + \sum_{\mu=\nu}^{\infty} (f_\mu - p_\mu) \quad \text{on} \quad U_\nu$$

$$= \psi_\nu + h_\nu,$$

where $\bar{\partial}h_\nu = 0$. So $\bar{\partial}\psi = \bar{\partial}\psi_\nu = \varphi$ for all ν.   Q.E.D.

COROLLARY. Let $A^{0,q}$ denote the sheaf of germs (over M) of C^∞ $(0,q)$-forms. Then $A^0 = A^{0,0} = \mathscr{D}$ and

$$0 \longrightarrow \mathcal{O} \overset{\bar{\partial}}{\longrightarrow} \mathscr{D} \overset{\bar{\partial}}{\longrightarrow} A^{0,1} \longrightarrow A^{0,2} \overset{\bar{\partial}}{\longrightarrow} \cdots \overset{\bar{\partial}}{\longrightarrow} A^{0,n} \overset{\bar{\partial}}{\longrightarrow} 0$$

is exact.

THEOREM 6.3.   (Dolbeault) $H^q(M, \mathcal{O}) \cong [H^0(M, \bar{\partial}A^{0,q-1})/\bar{\partial}H^0(M, A^{0,q-1})]$.

Proof. $A^{0,q}$ is a fine sheaf.   Q.E.D.

COROLLARY. $H^q(U_R, \mathcal{O}) = 0$ for $q \geq 1$.

Proof.   Use Theorems 6.2 and 6.3 $(H^0(U_R, \bar{\partial}A^{0,q-1}) = \bar{\partial}H^0(U_R, A^{0,q-1}))$.

Q.E.D.

We can generalize these results. Let φ be a $C^\infty(p,q)$-form. We have the following sequence of statements whose proofs are similar to the previous proofs [see Gunning and Rossi (1965)].

LEMMA 6.2′.   On U_R, a (p,q) form $\varphi, q \geq 1$ satisfies $\bar{\partial}\varphi = 0$ if and only if $\varphi = \bar{\partial}\psi$, where ψ is a $(p, q-1)$ form.

COROLLARY′.   Let Ω^p be the sheaf of germs of holomorphic p-forms over M.

Then

$$0 \longrightarrow \Omega^p \longrightarrow A^{p,0} \xrightarrow{\ \bar{\partial}\ } A^{p,1} \xrightarrow{\ \bar{\partial}\ } \cdots \xrightarrow{\ \bar{\partial}\ } A^{p,n} \longrightarrow 0$$

is a fine resolution of Ω^p.

THEOREM 6.3'. $H^q(M, \Omega^p) \cong [H^0(M, \bar{\partial}A^{p,q-1})/\bar{\partial}H^0(M, A^{p,q-1})]$.

REMARK. Let $\varphi^{(p,q)}$ be a (p,q)-form. Then $\overline{\bar{\partial}\varphi^{(p,q)}} = 0$ if and only if $\overline{\varphi^{(p,q)}} = \bar{\partial}\psi^{(p,q-1)}$. So,

$$\overline{\partial\varphi^{(p,q)}} = 0 \text{ if and only if } \overline{\varphi^{(p,q)}} = \partial\overline{\psi^{(p,q-1)}}.$$

Hence (on U_R) $\partial\varphi^{(p,q)} = 0$ if and only if $\varphi^{(p,q)} = \partial\psi^{(p-1,q)}$, $p \geq 1$.

If $f(z)$ is holomorphic on $|z| < R$, then $dg/dz = f$ where $g(z) = \int_0^z g(\zeta)\, d\zeta$.

LEMMA 6.3. Let φ be a holomorphic p-form on U_R, $p \geq 1$. Then there exists a holomorphic $(p-1)$-form ψ such that $d\psi = \varphi$ if and only if $d\varphi = 0$.

Proof. $d = \partial + \bar{\partial}$ so $d\varphi = \partial\varphi$ if φ is holomorphic and if $d\varphi = \partial\varphi = 0$. The rest of the proof uses the observation made just before the statement of the lemma. The details are essentially the same as in the previous proofs and are left to the reader.

THEOREM 6.4. $0 \longrightarrow \mathbb{C} \longrightarrow \mathcal{O} \xrightarrow{\ d\ } \Omega^1 \xrightarrow{\ d\ } \Omega^2 \longrightarrow \cdots \to \Omega^n \longrightarrow 0$
is exact.

REMARK. Ω^p is *not* a fine sheaf.

Now we consider holomorphic vector bundles F over M. Let F be defined by the 1-cocycle $\{f_{jk}(z)\}$ where $f_{jk}(z) = [f^\alpha_{jk\beta}(z)]_{\alpha,\beta=1,\ldots,m}$. On each coordinate patch $\mathcal{O}(F) = \mathcal{O} \oplus \cdots \oplus \mathcal{O}$ (m times). Let φ be a section of $\mathcal{O}(F)$ over an open set $W \subseteq M$. On $W \cap U_j$,

$$\varphi(z) = [\varphi_j^1(z), \cdots, \varphi_j^m(z)],$$

where φ_j^λ is a holomorphic function on $W \cap U_j$, and

$$\varphi_j^\lambda(z) = \sum_\mu f^\lambda_{jk\mu}(z)\varphi_k^\mu(z) \qquad \text{for } z \in W \cap U_j \cap U_k.$$

By a (p, q)-form φ with coefficients in F over W we mean $\varphi(z) = [\varphi_j^1(z), \cdots, \varphi_j^m(z)]$ where each $\varphi_j^\lambda(z)$ is a (p, q)-form over $W \cap U_j$ such that

$$\varphi_j^\lambda(z) = \sum_{\mu=1}^m f^\lambda_{jk}(z)\varphi_k^\mu(z)$$

as differential forms for $z \in W \cap U_j \cap U_k$. We define $A^{p,q}(F)$ to be the sheaf over M of germs of (p, q)-forms with coefficients in F. At each point $x \in M$, the stalk $A_x^{p,q}(F) = A_x^{p,q} \oplus \cdots \oplus A_x^{p,q}$ (m times). We have:

THEOREM 6.5. $0 \to \mathcal{O}(F) \to A^0(F) \xrightarrow{\ \bar{\partial}\ } A^{0,1}(F) \to \cdots \to \bar{\partial} A^{0,n} \to 0$ is a fine resolution of $\mathcal{O}(F)$.

Proof. We remark on the definition of $\bar{\partial}$ and leave the proof to the reader. The functions f_{jk} are holomorphic so $(\partial/\partial\bar{z})f_{jk}(z) = 0$. Hence,

$$\bar{\partial}\varphi_j^\lambda(z) = \sum_{\mu=1}^m f_{jk\mu}^\lambda(z)\, \bar{\partial}\varphi_k^\mu(z).$$

So $\bar{\partial}$ is well defined and if φ is a (p, q) form with coefficients in F, $\bar{\partial}\varphi$ is a $(p, q + 1)$-form with coefficients in F. We note that

$$\partial\varphi_j^\lambda(z) = \sum_\mu f_{jk\mu}^\lambda(z)\, \partial\varphi_k^\mu(z) + \sum_\mu \partial f_{jk\mu}^\lambda(z)\varphi_k^\mu(z),$$

and $\partial\varphi$ is *not* well defined on $A^{0,p}$. We also notice that $A^{p,q}(F) = A^{p,q}\otimes_{\mathcal{O}}\mathcal{O}(F)$. Remember that $\varphi^{(p,0)}$ is a differentiable section of $T^* \otimes \cdots \otimes T^*$ which is skew-symmetric. We define $T^* \wedge \cdots \wedge T^*$ to be the subbundle of $T^* \otimes \cdots \otimes T^*$ consisting of those $(z, \zeta_{\alpha_1 \cdots \alpha_p})$ which are skew-symmetric in $\alpha_1 \cdots \alpha_p$. Then $\varphi^{(p,0)}$ is a differentiable section of $(\wedge T^*)^p$, $\varphi^{(p,q)}$ is a differentiable section of $(\wedge T^*)^p \otimes (\bar{T}^*)^q$, $A^{(p,q)}$ is the sheaf of germs of differentiable sections of $(\wedge T^*)^p \otimes (\wedge \bar{T}^*)^q$, and $A^{(p,q)}(F)$ is the sheaf of germs of differentiable sections of $(\wedge T^*)^p \otimes (\wedge \bar{T}^*)^q \otimes F$.

[3]

Geometry of Complex Manifolds

1. Hermitian Metrics; Kähler Structures

Let M be a complex manifold. We want to introduce in M something analogous to a Riemannian metric which is "compatible" with the complex structure of M. Remember that for a Riemannian manifold the element of arc length is given by $ds^2 = \sum g_{\alpha\beta}\, dx^\alpha dx^\beta$.

DEFINITION 1.1. An *Hermitian metric* on a complex manifold M with local coordinates (z_j^ν) is given by

$$ds^2 = \sum_{\alpha,\beta=1}^{n} g_{j\alpha\bar\beta}(z)\, dz_j^\alpha\, d\bar z_j^\beta,$$

where $g_{j\alpha\bar\beta}(z)$ is a C^∞ section of $T^* \otimes \bar T^*$ such that

(1) $\overline{g_{j\alpha\bar\beta}(z)} = g_{j\beta\bar\alpha}(z)$ (Hermitian symmetric)

(2) $\sum_{\alpha,\beta=1}^{n} g_{j\alpha\bar\beta}(z)\, \zeta^\alpha\, \overline{\zeta^\beta} \geq 0$,   and equality if and only if $\zeta = 0$ (positive definite).

THEOREM 1.1. Given any complex manifold M, we can introduce an Hermitian metric on M.

Proof. Let $\mathscr{U} = \{U_j\}$ be a locally finite covering of M with coordinate patches U_j and let $(z_j^1, \cdots, z_j^n)$ be coordinates on U_j. On each U_j we have a metric

$$\sum_{\lambda=1}^{n} dz_j^\lambda\, d\bar z_j^\lambda = \sum_{\alpha,\beta=1}^{n} \delta_{\alpha\beta}\, dz^\alpha\, d\bar z^\beta,$$

the usual Euclidean metric. Let $\{\rho_j\}$ be a partition of unity subordinate to $\mathscr{U}$; that is,

$$\overline{\{z \mid \rho_j(z) > 0\}} \subset U_j.$$

Define

$$ds^2 = \sum_j \rho_j(z)\left(\sum_{\lambda=1}^{n} dz_j^\lambda\, d\bar z_j^\lambda \right).$$

83

We claim this is an Hermitian metric. On U_k

$$ds^2 = \sum_{\alpha,\beta=1}^{n} g_{k\alpha\bar\beta}(z)\, dz_k^\alpha\, d\bar z_k^\beta.$$

Since

$$dz_j^\lambda = \sum_\alpha \frac{\partial z_j^\lambda}{\partial z_k^\alpha}\, dz_k^\alpha,$$

$$g_{k\alpha\bar\beta}(z) = \sum_{j,\lambda} \rho_j(z)\left(\frac{\partial z_j^\lambda}{\partial z_k^\alpha}\right)\left(\overline{\frac{\partial z_j^\lambda}{\partial z_k^\beta}}\right).$$

Then

$$\sum_{\alpha,\beta,\lambda=1}^{n}\left(\frac{\partial z_j^\lambda}{\partial z_k^\alpha}\right)\left(\overline{\frac{\partial z_j^\lambda}{\partial z_k^\beta}}\right)\zeta^\alpha\bar\zeta^\beta = \left|\sum_{\alpha,\lambda=1}^{n}\frac{\partial z_j^\lambda}{\partial z_k^\alpha}\zeta^\alpha\right|^2 > 0$$

if $\zeta \neq 0$. Thus ds^2 is positive definite and the rest is easy to check.   Q.E.D.

Suppose given an Hermitian metric $ds^2 = \sum g_{j\alpha\bar\beta}\, dz_j^\alpha\, d\bar z_j^\beta$. Define $g_{j\lambda\mu}$ for $\lambda, \mu \in \{1, 2, \cdots, n, \bar 1, \cdots \bar n\}$ by

$$(g_{j\lambda\mu}) = \begin{pmatrix} 0 & g_{j\alpha\bar\beta} \\ g_{j\bar\alpha\beta} & 0 \end{pmatrix}.$$

Then $2\sum g_{j\alpha\bar\beta}\, dz_j^\alpha\, dz_j^{\bar\beta} = \sum_{\lambda,\mu \in \{1,2,\cdots,\bar n\}} g_{j\lambda\mu}\, dz_j^\lambda\, dz_j^\mu$, where, as has been our custom, $z_j^{\bar\alpha} = \bar z_j^\alpha$. We associate to ds^2 a differential form of type $(1,1)$, $\omega = i\sum g_{j\alpha\bar\beta}\, dz_j^\alpha \wedge dz_j^{\bar\beta}$, where $i = \sqrt{-1}$.

REMARK. $\bar\omega = -i\sum \overline{g_{j\alpha\bar\beta}}\, d\bar z_j^\alpha \wedge dz_j^\beta = i\sum g_{j\beta\bar\alpha}\, dz_j^\beta \wedge dz_j^{\bar\alpha} = \omega$. Or if one prefers

$$\overline{\omega(\zeta,\eta)} = \tfrac{1}{2}ig_{j\alpha\bar\beta}(\zeta_j^\alpha\eta_j^{\bar\beta} - \eta_j^\alpha\zeta_j^{\bar\beta})$$

$$= -\tfrac{1}{2}ig_{j\beta\bar\alpha}(\zeta_j^{\bar\alpha}\eta_j^\beta - \eta_j^{\bar\alpha}\zeta_j^\beta)$$

$$= \omega(\zeta,\eta).$$

So ω is a *real* form.

DEFINITION 1.2. $2\sum g_{\alpha\bar\beta}\, dz^\alpha\, dz^{\bar\beta}$ is called a *Kähler metric* if $d\omega = 0$. Then we call ω a *Kähler form*. M is called a *Kähler manifold* if we can define a Kähler metric on M.

REMARKS

(1) An Hermitian metric can always be defined, but generally $d\omega$ is not zero.

(2) Compact Kähler manifolds have many properties similar to (projective) algebraic manifolds.

(3) Every algebraic manifold is a Kähler manifold. (Recall that algebraic manifolds are by definition compact submanifolds of $\mathbb{P}^N$ for some N.) ·

(4) There are many nonalgebraic Kähler manifolds.

CONJECTURE. A compact complex manifold of complex dimension 2 is Kähler if and only if the first Betti number is even.

We know that M Kähler implies that the first Betti number is even (Theorem 5.4, Corollary 2). We have the following facts:

(1) If $b_1(M^2)$ is even, then M^2 is a deformation of an algebraic manifold where by M^2 we mean a compact complex manifold of complex dimension 2. Thus there is a complex family $\{M_t | \ |t| < 2\}$ such that M_0^2 is algebraic and $M_1^2 = M^2$ [Kodaira (1964)].

(2) Any *small* deformation of a compact Kähler manifold M_0^n is Kähler; that is, if M_0^n is Kähler, then M_t^n is Kähler for small enough t.

(3) It was conjectured that *any* deformation of a (compact) Kähler manifold is Kähler. This turned out to be false [H. Hironaka (1962)] for dimension ≥ 3. So we make the conjecture for M^2.

Now we collect some facts about Kähler manifolds. Let $U = \{z | \ |z| < 1\} \subset \mathbb{C}^n$.

PROPOSITION 1.1. A form $\varphi = \sum \varphi_{\alpha\bar\beta}\, dz^\alpha \wedge dz^\beta$ on U satisfies $d\varphi = 0$ if and only if there is a C^∞ function f such that

$$\varphi = \partial\bar\partial f = \partial\left(\sum \frac{\partial f}{\partial \bar z^\beta}\, dz^\beta\right) = \sum_{\alpha,\beta} \frac{\partial^2 f}{\partial \bar z^\alpha\, \partial \bar z^\beta}\, dz^\alpha \wedge dz^\beta;$$

that is, if and only if $\varphi_{\alpha\bar\beta} = (\partial^2 f / \partial z^\alpha\, \partial \bar z^\beta)$.

Proof

(1) If $\varphi = \partial\bar\partial f$, then $d(\partial\bar\partial f) = (\partial + \bar\partial)(\partial\bar\partial f)$

$$= \bar\partial\partial\bar\partial f$$
$$= -\partial\bar\partial\bar\partial f$$
$$= 0.$$

(2) Suppose $d\varphi = \partial\varphi + \bar\partial\varphi = 0$. Since $\partial\varphi$ is of type $(2, 1)$ and $\bar\partial\varphi$ is of type $(1, 2)$, $\bar\partial\varphi = 0$ and $\partial\varphi = 0$. Dolbeault's lemma implies that there is $\psi^{(1,0)}$ such that $\bar\partial\psi^{(1,0)} = \varphi$ on U, $\psi^{(1,0)} = \sum \psi_\alpha\, dz_\alpha$. We have

$$0 = \partial\varphi = \partial\bar\partial\psi^{(1,0)} = -\bar\partial\partial\psi^{(1,0)}$$

and

$$\partial \psi^{(1,0)} = \frac{1}{2} \sum \left(\frac{\partial \psi_\beta}{\partial z^\alpha} - \frac{\partial \psi_\alpha}{\partial z^\beta} \right) dz^\alpha \wedge dz^\beta.$$

So

$$\frac{\partial}{\partial \bar{z}^\lambda} \left(\frac{\partial \psi_\beta}{\partial z^\alpha} - \frac{\partial \psi_\alpha}{\partial z^\beta} \right) = 0$$

and $\partial \psi^{(1,0)}$ is a holomorphic 2-form. According to Chapter 2, Lemma 6.3, there is a holomorphic 1-form η on U such that $\partial \psi^{(1,0)} = d\eta = \partial \eta$. Thus $\partial[\psi^{(1,0)} - \eta] = 0$ and there is a C^∞ function f such that $\psi^{(1,0)} - \eta = \partial f$ (see the remark after Chapter 2, Theorem 6.3′). Hence,

$$\varphi = \bar{\partial} \psi^{(1,0)} = \bar{\partial}(\eta + \partial f) = \bar{\partial}\partial f = \partial \bar{\partial}(-f). \qquad \text{Q.E.D.}$$

Let $2 \sum g_{j\alpha\beta} \, dz_j^\alpha \, d\bar{z}_j^\beta$ be an Hermitian metric and $\omega = i \sum g_{j\alpha\bar{\beta}} \, dz_j^\alpha \wedge d\bar{z}_j^\beta$ be the associated form on U_j where the manifold $M = \cup U_j$. Assume $U_j = \{z_j \mid |z_j| < 1\}$.

THEOREM 1.2. ω is a Kähler form (that is, $d\omega = 0$) if and only if there is a differentiable function K_j on each U_j such that $\omega = \partial \bar{\partial} K_j$ on U_j.

Proof. Use Proposition 1.1.

PROPOSITION 1.2. $\mathbb{P}^n$ is a Kähler manifold.

Proof. $\mathbb{P}^n = \bigcup_{j=0}^n (U_j)$, where $U_j = \{\zeta \mid \zeta = (\zeta_0, \cdots, \zeta_n), \zeta_j \neq 0\}$. On U_j then we have (affine) coordinates $z_j = (z_j^0, \cdots, z_j^{j-1}, z_j^{j+1}, \cdots, z_j^n)$, where $z_j^\alpha = \zeta_\alpha / \zeta_j$. We set

$$K_j = \log\left(1 + \sum_{\alpha \neq j} |z_j^\alpha|^2\right)$$

$$= \log\left(\sum_{\alpha=0}^n |\zeta_\alpha|^2\right) - \log|\zeta_j|^2$$

on U_j. Then

$$K_j - K_k = \log|\zeta_k/\zeta_j|^2 = \log|z_j^k|^2$$

$$= \log z_j^k + \log \overline{z_j^k} \text{ on } U_j \cap U_k.$$

Therefore $\partial \bar{\partial} K_j = \partial \bar{\partial} K_k$ on $U_j \cap U_k$, and we can define a global form $\omega = i \, \partial \bar{\partial} K_j$ on each U_j. Clearly $d\omega = 0$. If $\omega = i \sum g_{j\alpha\bar{\beta}} \, dz_j^\alpha \wedge d\bar{z}_j^\beta$, we must show that $g_{j\alpha\bar{\beta}}$ is positive definite; Hermitian symmetric will be left to the reader. Take $j = 0$ and let $z^\alpha = z_0^\alpha$. Then

$$K_0 = \log\left(1 + \sum_{\alpha=1}^{n} |z^\alpha|^2\right),$$

so

$$\bar\partial K_0 = \frac{\sum z^\alpha \, d\bar z^\alpha}{1 + \sum |z^\alpha|^2}$$

and

$$\partial\bar\partial K_0 = \frac{\sum dz^\alpha \wedge d\bar z^\alpha}{1 + \sum |z^\alpha|^2} - \frac{\sum \bar z^\beta dz^\beta \wedge \sum z^\alpha \, d\bar z^\alpha}{(1 + \sum |z^\alpha|^2)^2}$$

$$= \frac{1}{(1 + \sum |z^\alpha|^2)^2}\left(\delta_{\alpha\beta}(1 + \sum |z_\alpha|^2) - \bar z^\alpha z^\beta\right) dz^\alpha \wedge d\bar z^\beta.$$

Then

$$\sum (\delta_{\alpha\beta}(1 + \sum |z_\alpha|^2) - \bar z^\alpha z^\beta)\zeta^\alpha \bar\zeta^\beta = (\zeta, \zeta)^2(1 + (z, z)^2) - |(\zeta, z)|^2, \quad (1)$$

where

$$(\zeta, z) = \sum_{\alpha=1}^{n} \zeta^\alpha \bar z^\alpha.$$

The Schwarz inequality implies that $|(\zeta, z)|^2 \le (\zeta, \zeta)^2(z, z)|^2$ hence Equation (1) is a positive definite form.   Q.E.D.

THEOREM 1.3.   Any submanifold of a Kähler manifold is a Kähler manifold.

Proof. Suppose that $N \subset M$ and ω is a Kähler form on M. We claim that the restricted form $\omega|_N$, which we shall shortly describe, is a Kähler form on N. To define the restriction we proceed as follows: On each coordinate patch U_j with $U_j \cap N \ne \phi$ we choose local coordinates $(z_j^1, \cdots, z_j^n, \cdots, z_j^m) = (z_j^1, \cdots, z_j^n, w_j^1, \cdots, w_j^r)$ with $r = m - n$, $m = \dim M$, $n = \dim N$ so that

$$N \cap U_j = \{(z_j^1, \cdots, z_j^n, 0, \cdots, 0)\}.$$

Suppose

$$\varphi = \frac{1}{p!} \sum \varphi_{j\alpha_1 \cdots \alpha_p}(z_j, w_j) \, dz^{\alpha_1} \wedge \cdots \wedge dz_j^{\alpha_p}$$

$$+ \sum \varphi_{j\alpha_1 \cdots \alpha_p}(z_j, w_j) \, dz_j^{\alpha_1} \wedge \cdots \wedge dw_j^{\alpha_q} \wedge \cdots \wedge dw_j^{\alpha_p}, \quad (2)$$

then

$$\varphi \mid N = \frac{1}{p!} \sum \varphi_{j\alpha_1 \cdots \alpha_p}(z_j, 0) \, dz_j^{\alpha_1} \wedge \cdots \wedge dz_j^{\alpha_p} \qquad (3)$$

is a form on N since

$$z_j^\alpha = f_{jk}^\alpha(z_k, w_k), \qquad w_j^\lambda = g_{jk}^\lambda(z_k, w_k),$$

$$0 = g_{jk}^\lambda(z_k, 0),$$

and on N,

$$d\omega_j^\lambda = \sum \frac{\partial g_{jk}^\lambda}{\partial w_k^\mu}\, dw_k^\mu.$$

Similar remarks apply to forms of type (p,q). We see that $\varphi | N$ is a form of type (p,q) and that

$$d(\varphi \mid N) = d\varphi \mid N.$$

Thus we see easily that $\omega \mid N$ is positive definite and $d(\omega \mid N) = 0$.   Q.E.D.

COROLLARY.   Any algebraic manifold M is a Kähler manifold.

Proof. $M \subseteq \mathbb{P}^n$ for some n.

THEOREM 1.4.   If M is compact and Kähler, then the Betti numbers $b_{2k}(M) \geq 1$ for $k = 1, \cdots, n$, $n = \dim_{\mathbb{C}} M$.

Before giving the proof we shall embark on a brief discussion of integrals of forms, specifically m-forms, on a compact differentiable manifold M of dimension m. Suppose $x_j = (x_j^1, \cdots, x_j^m)$ is a local coordinate on $U_j \subset M$. Then an m-form φ can be written

$$\varphi = \varphi_{j12\ldots m}(x)\, dx_j^1 \wedge \cdots \wedge dx_j^m \text{ on } U_j$$

$$= \varphi_{k_1\ldots m}(x)\, dx_k^1 \wedge \cdots \wedge dx_k^m \text{ on } U_k, \text{ and on } U_j \cap U_k,$$

$$dx_j^1 \wedge \cdots \wedge dx_j^m = \det\left(\frac{\partial x_j^\alpha}{\partial x_k^\beta}\right) dx_k^1 \wedge \cdots \wedge dx_k^m. \tag{4}$$

To be able to define $\int_M \varphi$ we make the assumption that M is *oriented*. This means that there is a covering of M with local coordinate patches (U_j, x_j) such that $\det(\partial x_j^\alpha/\partial x_k^\beta) > 0$ on $U_j \cap U_k$. Any local coordinate (X_ν, ξ_ν) is *positively oriented* if $\det(\partial \xi_\nu^\alpha/\partial x_j^\beta) > 0$ for all j such that $U_j \cap X_\nu \neq \phi$. Let $\{\rho_j\}$ be a partition of unity subordinate to $\{U_j\}$.

DEFINITION 1.3. $\int_M \varphi = \sum_j \int_{U_j} \rho_j(x)\varphi_{j_1\ldots m}(x)\, dx_j^1 \cdots dx_j^m$.   Using Equation (4) and the orientability it is easy to check that $\int_M \varphi$ is well defined independent of the choice of covering by local (positively oriented) coordinate patches and partition of unity subordinate to it. We leave this to the reader.

PROPOSITION 1.3.   If $\varphi = d\psi$, then $\int_M \varphi = 0$.

Proof. $\psi = \sum_j \rho_j(x) \, \psi_j(x)$, so

$$\int_M \varphi = \int \sum_j d(\rho_j \psi_j) = \sum_j \int_{U_j} d(\rho_j \psi_j),$$

where

$$U_j = \{x_j | \, |x_j^\alpha| < r_j\} \text{ and } \rho_j(x) = 0 \text{ for } x \notin \overline{W}_j \subseteq U_j.$$

We will show that each $\int_{U_j} d(\rho_j \psi_j) = 0$. To simplify notation let us drop the subscripts. Then

$$\rho_j \psi_j(x) = \sum_{\alpha=1}^m h_\alpha(x) \, dx^1 \wedge \cdots dx^{\alpha-1} \wedge dx^{\alpha+1} \wedge \cdots \wedge dx^m$$

and $h_\alpha(x) = 0$ for $x \notin \overline{W}_j$. We calculate $\int_{U_j} d(\rho_j \psi_j)$:

$$\int_{U_j} d(\rho_j \psi_j) = \int_{U_j} d\left(\sum_\alpha h_\alpha \, dx^1 \wedge \cdots \wedge dx^{\alpha-1} \wedge dx^{\alpha+1} \wedge \cdots \wedge dx^m \right)$$

$$= \int_{U_j} \sum_\alpha (-1)^{\alpha+1} \frac{\partial h_\alpha}{\partial x_\alpha} \, dx^1 \cdots dx^m$$

$$= \sum_\alpha (-1)^{\alpha+1} \int_{|x^i|<r} dx^1 \cdots dx^{\alpha-1} \, dx^{\alpha+1} \cdots dx^m \int_{-r}^r \frac{\partial h_\alpha}{\partial r^\alpha} \, dx^\alpha$$

$$= 0,$$

since

$$\int_{-r}^r \frac{\partial h^\alpha}{\partial x^\alpha} \, dx^\alpha = h_\alpha(r) - h_\alpha(-r) = 0. \qquad \text{Q.E.D.}$$

A complex manifold M is naturally oriented. For if $(z_j^1, \cdots, z_j^n)$ are local coordinates and $z_j^\alpha = x_j^{2\alpha-1} + i \, x_j^{2\alpha}$, then the coordinates $(x_j^1, \cdots, x_j^{2n})$ give a covering of M and the determinant of a change of charts is $\det |\partial z_j^\alpha / \partial z_k^\beta|^2 > 0$ hence M is oriented and $(x_j^1, \cdots, x_j^{2n})$ is a positively oriented chart.

Proof.   (of Theorem 1.4)   Let $\omega = i \sum g_{\alpha\bar\beta} \, dz^\alpha \wedge dz^{\bar\beta}$. Then

$$\omega^n = \omega \wedge \cdots \wedge \omega$$

$$= (i)^n \sum g_{\alpha_1\bar\beta_1} \, g_{\alpha_2\bar\beta_2} \cdots g_{\alpha_n\bar\beta_n} \, dz^{\alpha_1} \wedge dz^{\bar\beta_1} \wedge \cdots \wedge dz^{\alpha_n} \wedge dz^{\bar\beta_n}.$$

We claim $\int_M \omega^n > 0$. For

$$\omega^n = (i)^n \sum_{\alpha,\beta} \mathrm{sgn}\begin{pmatrix} 1 & \cdots & n \\ \alpha_1 & \cdots & \alpha_n \end{pmatrix} \mathrm{sgn}\begin{pmatrix} 1 & \cdots & n \\ \beta_1 & \cdots & \beta_n \end{pmatrix} g_{\alpha_1\bar{\beta}_1} \cdots$$

$$g_{\alpha_n\bar{\beta}_n}\, dz^1 \wedge dz^{\bar{1}} \cdots \wedge dz^n \wedge dz^{\bar{n}},$$

and

$$\sum_{\alpha_1,\cdots,\alpha_n} \mathrm{sgn}\begin{pmatrix} 1 & \cdots & n \\ \alpha_1 & \cdots & \alpha_n \end{pmatrix} g_{\alpha_1\bar{\beta}_1} \cdots g_{\alpha_n\bar{\beta}_n} = \begin{vmatrix} g_{1\bar{\beta}_1} & \cdots & g_{1\bar{\beta}_n} \\ \vdots & & \vdots \\ g_{n\bar{\beta}_1} & \cdots & g_{n\bar{\beta}_n} \end{vmatrix}$$

$$= \mathrm{sgn}\begin{pmatrix} 1 & \cdots & n \\ \beta_1 & \cdots & \beta_n \end{pmatrix} \begin{vmatrix} g_{1\bar{1}} & \cdots & g_{1\bar{n}} \\ \vdots & & \vdots \\ g_{n\bar{1}} & \cdots & g_{n\bar{n}} \end{vmatrix}$$

$$= \mathrm{sgn}\begin{pmatrix} 1 & \cdots & n \\ \beta_1 & \cdots & \beta_n \end{pmatrix} g,$$

where $g = \det(g_{\alpha\bar{\beta}})$. Therefore,

$$\omega^n = (i)^n n!\, g\, dz^1 \wedge \cdots \wedge dz^{\bar{n}}.$$

Since

$$dz^1 \wedge dz^{\bar{1}} = (dx^1 + i\,dx^2) \wedge (dx^1 - i\,dx^2)$$
$$= -2\,i\,dx^1 \wedge dx^2,$$
$$\omega^n = 2^n\, n!\, g\, dx^1 \wedge \cdots \wedge dx^{2n}.$$

We have assumed $(g_{\alpha\bar{\beta}})$ is positive definite so $g > 0$. Thus $\int_M \omega^n > 0$. If $\omega^n = d\psi$ then $\int_M \omega^n = 0$ by Proposition 1.3. Thus $\omega^n \neq d\psi$. In fact, we claim $\omega \wedge \cdots \wedge \omega = \omega^k \neq d\psi$ for any ψ. Since $d\omega = 0$, $d(\omega^k) = 0$ and if $\omega^k = d\psi$ then $\omega^n = \omega^k \wedge \omega^{n-k} = d\psi \wedge \omega^{n-k} = d(\psi \wedge \omega^{n-k})$. Now recall that

$$b_{2k} = \dim_{\mathbb{C}} H^{2k}(M, \mathbb{C}) = \dim_{\mathbb{C}} \frac{H^0(M, dA^{2k-1})}{dH^0(M, A^{2k-1})}.$$

The facts just proved show that

$$\omega^k \in H^0(M, dA^{2k-1}),$$
$$\omega^k \notin dH^0(M, A^{2k-1}).$$

Thus, $b_{2k} \geq 1$.

THEOREM 1.5. If M is compact Kähler and if $N \subset M$ is a compact complex submanifold, then N is not homologous to zero in M.

Proof. (sketch) We first prove:

PROPOSITION 1.4.   (Stokes' theorem)   If W^{n+1} is a compact differentiable manifold with boundary $\partial W^{n+1} = M^n$, then $\int_{W^{n+1}} d\psi = \int_{M^n} \psi$ for any n-form ψ on $W = W^{n+1}$.

Proof. We cover W with a locally finite family of coordinate patches $\{U_j\}$ such that $U_j = \{x_j \mid |x_j^\alpha| < r_j\}$ if $U_j \subseteq \text{int}(W)$ and if $U_j \cap M \neq \phi$, then $U_j = \{x_j \mid |x_j^\alpha| < r_j, \ \alpha > 1, \ -r_j < x_j^1 \leq r_j\}$, and $M \cap U_j = \{x_j \mid x_j^1 = r_j\}$. We choose a partition of unity $\{\rho_j\}$ with the following properties:

(1)   supp $\rho_j = \overline{\{x \mid \rho_j(x) > 0\}} \subset U_j$ if $U_j \cap M = \phi$.
(2)   supp $\rho_j \subseteq \{x_j \mid -r_j < x_j^1 \leq r_j, \ |x_j^\alpha| < r_j, \ \alpha \geq 2\}$.
(3) ρ_j is C^∞, $\rho_j \geq 0$ and $\sum \rho_j = 1$.

Then $\int_W d\psi = \sum_j \int_W d(\rho_j \psi) = \sum_j \int_{U_j} d(\rho_j \psi)$. From Proposition 1.3 $\int_{U_j} d(\rho_k \psi) = 0$ if $U_k \cap M = \phi$. Suppose $U_j \cap M \neq \phi$ and $\psi = \sum_{\alpha=1}^{n+1} \psi_j^\alpha \, dx_j^1, \wedge \cdots dx_j^{\alpha-1} \wedge dx^{\alpha+1} \wedge \cdots \wedge dx_j^{n+1}$ on U_j. Then

$$\int_{U_j} d(\rho_j x) = \int_{U_j} (-1)^{\alpha+1} \frac{\partial(\rho_j \psi_j^\alpha)}{\partial x_j^\alpha} \, dx_j^1 \cdots dx_j^{n+1}$$

and

$$\int_{-r_j}^{r_j} \frac{\partial(\rho_j \psi^\alpha)}{\partial x_j^\alpha} \, dx_j^\alpha = 0 \text{ if } \alpha \neq 1.$$

For $\alpha = 1$,

$$\int_{-r_j}^{r_j} \frac{\partial(\rho_j \psi^\alpha)}{\partial x_j^\alpha} \, dx_j^\alpha = \rho_j \psi_j^1(r_j, x_j^2, \cdots, x_j^{n+1}).$$

Hence,

$$\int_{U_j} d(\rho_j \psi) = \int_{U_j} \rho_j \psi_j^1(r_j, x_j^2, \cdots, x_j^{n+1}) \, dx_j^2 \cdots dx_j^{n+1}$$

$$= \int_{M \cap U_j} \rho_j \psi.$$

Thus,

$$\int_W d\psi = \sum_j \int_{U_j \cap M} \rho_j \psi = \int_M \sum \rho_j \psi = \int_M \psi,$$

where $M = \partial W$.

For the proof of the theorem, suppose M^n is compact Kähler and N^m is a complex submanifold $N^m \subset M^n$. Suppose $N = \partial W \subset M$ for W an embedded submanifold. Then if ω is the Kähler form on M, $0 < \int_N \omega^m = \int_W d\omega^m = 0$.

This contradiction proves the theorem in this case. Generally if N is homologous to zero, we do not have such a convenient situation. One must change the proof, and we supply no details here.

2.   Norms and Dual Forms

Let Ω^p be the sheaf over M (a compact complex manifold) of holomorphic p-forms. Let $A^{p,q}$ be the sheaf of C^∞ (p, q)-forms on M. We want to introduce an Hermitian scalar product (φ, ψ) for $\varphi, \psi \in \Gamma(A^{p,q})$, which makes $\Gamma(A^{p,q})$ into an (incomplete) inner product space.

We introduce an Hermitian metric $2 \sum g_{j\alpha\bar\beta} \, dz_j^\alpha \, dz_j^{\bar\beta} = 2 \sum g_{\alpha\bar\beta} \, dz^\alpha \, dz^{\bar\beta}$ on M. Associated to this metric we have the form $\omega = i \sum g_{\alpha\bar\beta} \, dz^\alpha \wedge dz^{\bar\beta}$ and $\omega^n = 2^n \, n! \, g \, dx^1 \wedge \cdots \wedge dx^{2n}$ as before where $x^{2\alpha-1} + i \, x^{2\alpha} = z^\alpha$ and $g = \det(g_{\alpha\bar\beta})$. We denote the inverse $(g_{\alpha\bar\beta})^{-1}$ of $(g_{\alpha\bar\beta})$ by $(g^{\bar\alpha\beta}) = (g_{\alpha\bar\beta})^{-1}$, that is,

$$\sum_\beta g^{\bar\alpha\beta} g_{\beta\bar\gamma} = \delta_{\bar\gamma}^{\bar\alpha} \quad \text{and} \quad \sum_\beta g_{\alpha\bar\beta} g^{\bar\beta\gamma} = \delta_\alpha^\gamma.$$

The length $|\zeta|$ of a tangent vector ζ is given by $|\zeta^2| = \sum_{\alpha,\beta} g_{\alpha\bar\beta} \zeta^\alpha \bar\zeta^\beta$ and the inner product $(\zeta, \eta) = \sum_{\alpha,\beta} g_{\alpha\bar\beta} \zeta^\alpha \bar\eta^\beta$. Let

$$\varphi(z) = \frac{1}{p! \, q!} \sum \varphi_{\alpha_1 \cdots \bar\beta_q} \, dz^{\alpha_1} \wedge \cdots \wedge dz^{\bar\beta_q}$$

and

$$\psi(z) = \frac{1}{p! \, q!} \sum \psi_{\alpha_1 \cdots \bar\beta_q} \, dz^{\alpha_1} \wedge \cdots \wedge dz^{\bar\beta_q}.$$

Then at each point $z \in M$ we define

$$(\varphi, \psi)(z) = \frac{1}{p! \, q!} \sum_{\alpha, \beta, \lambda, \mu} g^{\bar\lambda_1\alpha_1} \cdots g^{\bar\beta_1\mu_1} \cdots g^{\bar\beta_q\mu_q} \varphi_{\alpha_1 \cdots \bar\beta_q} \overline{\psi_{\lambda_1 \cdots \bar\mu_q}}. \tag{1}$$

DEFINITION 2.1.   The *inner product* of two forms φ, ψ is

$$(\varphi, \psi) = \int_M (\varphi, \psi)(z) \, \frac{\omega^n}{n!} = \int_M (\varphi, \psi)(z) \, 2^n g \, dx^1 \cdots dx^{2n}. \tag{2}$$

$(,)$ satisfies the following properties:

    (1) $(\varphi, \psi) = \overline{(\psi, \varphi)}$,
    (2) $(a\psi + b\chi, \varphi) = a(\psi, \varphi) + b(\chi, \varphi)$,
    (3) $(\varphi, \varphi) \geq 0$,
    (4) $(\varphi, \varphi) = 0$ if and only if $\varphi = 0$.

We define $\|\varphi\| = \sqrt{(\varphi, \varphi)}$ as usual.

THEOREM 2.1.   There is a linear map $*\colon \Gamma(A^{p,q}) \to \Gamma(A^{n-p,n-q})$ such that:

(1) $(\varphi, \psi)(z) \dfrac{\omega^n}{n!} = \varphi(z) \wedge *\bar{\psi}(z),$

(2) $\overline{*\psi} = *\bar{\psi}$ (that is, $*$ is a real operator),

(3) $**\psi^{(p,q)} = (-1)^{p+q} \psi^{(p,q)}.$

Proof.   Before giving the proof let us fix some notation. Let $n = \dim M$. We denote as follows:

$$A_p = (\alpha_1, \cdots, \alpha_p), \alpha_1 < \alpha_2 < \cdots < \alpha_p, 1 \le \alpha_i \le n,$$

$$A_{n-p} = (\alpha_{p+1}, \cdots, \alpha_n), \alpha_{p+1} < \cdots < \alpha_n, 1 \le \alpha_i \le n$$

and $(\alpha_1, \cdots, a_p, \alpha_{p+1}, \cdots, \alpha_n)$ is a permutation of $(1, \cdots, n)$. For example, $n = 5$ and $A_2 = (2, 4)$, $A_{n-2} = (1, 3, 5)$. Similarly for $B_q = (\beta_1, \cdots, \beta_q)$, $B_{n-q} = (\beta_{q+1}, \cdots, \beta_n)$. Then with this notation we write a (p, q)-form

$$\psi = \sum_{A_p} \sum_{B_q} \psi_{A_p \bar{B}_q} \, dz^{A_p} \wedge dz^{\bar{B}_q}, \tag{3}$$

where $dz^{A_p} = dz^{\alpha_1} \wedge \cdots \wedge dz^{\alpha_p}$, and so on. We denote

$$\psi^{A_p \bar{B}_q} = \sum_{\lambda, \mu} g^{\bar{\alpha}_1 \lambda_1} \cdots g^{\bar{\mu}_q \beta_q} \psi_{\lambda_1 \cdots \lambda_p \bar{\mu}_1 \cdots \bar{\mu}_q}.$$

Then

$$\bar{\psi} = \sum_{A_p} \sum_{B_q} \overline{\psi_{A_p \bar{B}_q}} \, dz^{\bar{A}_p} \wedge dz^{B_q}$$

$$= \sum \sum (-1)^{pq} \overline{\psi_{A_p \bar{B}_q}} \, dz^{B_q} \wedge dz^{\bar{A}_p}$$

$$= \sum \sum (\bar{\psi})_{B_q \bar{A}_q} \, dz^{B_q} \wedge dz^{\bar{A}_q}.$$

Thus,

$$(\bar{\psi})_{\bar{B}_q \bar{A}_p} = (-1)^{pq} \overline{\psi_{A_p \bar{B}_q}}.$$

We can now write Equation (1) as

$$(\varphi, \psi)(z) = \sum_{A_p, B_q} \varphi_{A_p \bar{B}_q} \overline{\psi^{\bar{A}_p B_q}} = (-1)^{pq} \sum_{A_p, B_q} \varphi_{A_p B_q} \bar{\psi}^{\bar{B}_q A_p}. \tag{4}$$

Remember

$$g_{A_p A_{n-p} B_q B_{n-q}} = g_{\alpha_1 \cdots \alpha_n \bar{\beta}_1 \cdots \bar{\beta}_n} = \det(g_{\alpha_i \bar{\beta}_k}).$$

Then we define

$$*\psi = (i)^n (-1)^{\frac{1}{2}(n)(n-1)+pn} \sum_{A_q} \sum_{B_p} g_{A_q A_{n-q} B_p B_{n-p}} \psi^{B_p A_q} \, dz^{A_{n-q}} \wedge dz^{B_{n-p}}.$$

First we prove Theorem 2.1(1); that is,

$$\varphi \wedge *\overline{\psi} = (\varphi, \psi)(z)\,\frac{\omega^n}{n!}.$$

If ψ is as in Equation (3),

$$*\overline{\psi} = (i)^n(-1)^{\frac{1}{2}n(n-1)+qn} \sum_{A_p} \sum_{B_q} g_{A_p A_{n-p} B_q B_{n-q}} \overline{\psi}^{B_q A_p}\, dz^{A_{n-p}} \wedge dz^{\overline{B}_{n-q}}. \tag{5}$$

Let

$$\varphi = \sum_{\Lambda_p} \sum_{N_q} \varphi_{\Lambda_p \overline{N}_q}\, dz^{\Lambda_p} \wedge dz^{\overline{N}_q},$$

where $\Lambda_p = (\lambda_1, \cdots, \lambda_p)$, $N_q = (\nu_1, \cdots, \nu_q)$ with the usual conventions. Then

$$\varphi \wedge *\overline{\psi} = (\pm i^{\bar{n}}) \sum g_{A_p A_{n-p} B_q B_{n-q}} \overline{\psi}^{B_q A_p} \varphi_{\Lambda_p \overline{N}_q}\, dz^{\Lambda_p} \wedge dz^{\overline{N}_q} \wedge dz^{A_{n-p}} \wedge dz^{\overline{B}_{n-q}}$$

$$= (\pm i^n)(-1)^{q(n-p)} \sum (\cdots)\, dz^{\Lambda_p} \wedge dz^{A_{n-p}} \wedge dz^{\overline{N}_q} \wedge dz^{\overline{B}_{n-q}},$$

where $(\pm i^n) = i^n(-1)^{\frac{1}{2}(n-1)n+qn}$ from (4). Now, $dz^{\Lambda_p} \wedge dz^{A_{n-p}} \neq 0$ if and only if $\Lambda_p = A_p$; similarly we need only consider $N_q = B_q$. Thus

$$\varphi \wedge *\overline{\psi} = i^n(-1)^{\frac{1}{2}n(n-1)-qp} \sum \varphi_{A_p \overline{B}_q} \overline{\psi}^{B_q A_p} g_{A_p A_{n-p} B_q B_{n-q}}\, dz^{A_p A_{n-p}} \wedge dz^{\overline{B_q B_{n-q}}}.$$

Recall as in the proof of Theorem 1.4

$$g_{A_n A_{n-p} B_q B_{n-q}} = \operatorname{sgn} \alpha \cdot \operatorname{sgn} \beta \cdot g,$$

$$dz^{A_p A_{n-p}} = \operatorname{sgn} \alpha\, dz^1 \wedge \cdots \wedge dz^n,$$

$$dz^{\overline{B_q B_{n-q}}} = \operatorname{sgn} \beta\, dz^{\overline{1}} \wedge \cdots \wedge dz^{\overline{n}},$$

where $\operatorname{sgn} \alpha = \operatorname{sgn}\begin{pmatrix} 1 & \cdots & n \\ \alpha_1 & \cdots & \alpha_n \end{pmatrix}$ and so on, and

$$dz^1 \wedge \cdots \wedge dz^n \wedge dz^{\overline{1}} \wedge \cdots \wedge dz^{\overline{n}}$$

$$= (-1)^{n(n-1)/2}\, dz^1 \wedge dz^{\overline{1}} \wedge \cdots \wedge dz^n \wedge dz^{\overline{n}}.$$

Thus,

$$\varphi \wedge *\overline{\psi} = i^n(-1)^{pq} \sum \varphi_{A_p \overline{B}_q} \overline{\psi}^{B_q A_p} g\, dz^1 \wedge dz^{\overline{1}} \wedge \cdots \wedge dz^{\overline{n}}$$

$$= 2^n(\varphi, \psi)(z) g\, dx^1 \wedge \cdots \wedge dx^{2n}\,[\text{by Equation (4)}]$$

$$= (\varphi, \psi)(z) \cdot \frac{\omega^n}{n!}\, [\text{by Section 1, Equation (5)}].$$

Thus Theorem 2.1(1) is proved. For Theorem 2.1(2),

$$\overline{*\psi} = (-i)^n(-1)^{n(n-1)/2+pn} \sum g_{B_p B_{n-p} \overline{A}_q \overline{A}_{n-q}} \overline{\overline{\psi}^{B_p A_q}}\, dz^{\overline{A}_{n-q}} \wedge dz^{B_{n-p}}$$

$$= *\overline{\psi}$$

since

$$\tfrac{1}{2}n(n-1) + pn + n + (n-q)(n-p) + pq = \tfrac{1}{2}n(n-1) + n^2 + n - nq$$
$$\equiv \tfrac{1}{2}n(n-1) + nq \pmod 2.$$

Last we must check Theorem 2.1(3). Before doing this we make an assumption which we could have made for 2.1(1) and 2.1(2), which simplifies the calculations. We must check 2.1(3) pointwise and at any point z_0. We may assume by a change of coordinates that $g_{\alpha\bar\beta}(z_0) = \delta_{\alpha\bar\beta}$. This will *not* be true in a neighborhood of z_0; we only assume it at z_0. But since we only check 2.1(3) pointwise, each time we verify it at a point we may assume that $g_{\alpha\bar\beta} = \delta_{\alpha\bar\beta}$. Then

$$g_{A_q A_{n-q} B_p B_{n-p}}(z_0) = \operatorname{sgn}\begin{pmatrix} \alpha_1 & \cdots & \alpha_n \\ \beta_1 & \cdots & \beta_n \end{pmatrix}$$
$$= \operatorname{sgn}\begin{pmatrix} \alpha \\ \beta \end{pmatrix},$$

and $\psi^{\bar{B}_p A_q} = \psi_{B_p \bar{A}_q}$. Thus

$$(*\psi)_{A_{n-q} B_{n-p}} = (i)^n (-1)^{\frac{1}{2}n(n-1)+np} \operatorname{sgn}\begin{pmatrix} \alpha \\ \beta \end{pmatrix} \psi_{B_p \bar{A}_q}$$

at z_0 and

$$*(*\psi)_{B_p \bar{A}_q} = (i)^n (-1)^{\frac{1}{2}n(n-1)+n(n-q)} \operatorname{sgn}\begin{pmatrix} B_{n-p} & B_p \\ A_{n-q} & A_q \end{pmatrix} (*\psi)_{A_{n-q} \bar{B}_{n-p}}$$
$$= (-1)^{n + n(n-q) + q(n-q) + p(n-p) + np} \psi_{B_p \bar{A}_q}$$
$$= (-1)^{p+q} \psi_{B_q \bar{A}_q}.\qquad \text{Q.E.D.}$$

$*$ has one more property:

PROPOSITION 2.1. $\bar\varphi \wedge *\psi = \psi \wedge *\bar\varphi$ for $\varphi,\ \psi \in \Gamma(A^{p,q})$.

Proof. $\overline{\varphi \wedge *\psi}(z) = \overline{(\varphi,\psi)(z)}\omega^n/n! = (\psi,\varphi)(z)\omega^n/n!$
$$= \psi \wedge *\bar\varphi(z).\quad \text{(Remember that } \omega^n \text{ is real.)}$$

But also, $\overline{\varphi \wedge *\bar\psi} = \bar\varphi \wedge *\psi.\qquad \text{Q.E.D.}$

We next define the adjoints of $\bar\partial$, ∂, and d.

DEFINITION 2.2. $\vartheta\psi = -*\partial(*\psi)$, $\bar\vartheta\psi = -*(\bar\partial * \psi)$, and $\vartheta\psi = -*d(*\psi)$.

It is easy to see that

$$\vartheta: \Gamma(A^{(p,q)}) \longrightarrow \Gamma(A^{(p,q-1)}),$$

$$\bar{\vartheta}: \Gamma(A^{(p,q)}) \longrightarrow \Gamma(A^{(p-1,q)}),$$

$$\delta: \Gamma(A^p) \longrightarrow \Gamma(A^{p-1}).$$

PROPOSITION 2.2. Assume M is compact. Then

$$(\bar{\partial}\varphi, \psi) = (\varphi, \vartheta\psi) \text{ for } \varphi \in \Gamma(A^{(p,q-1)}), \psi \in \Gamma(A^{(p,q)}),$$

$$(\partial\varphi, \psi) = (\varphi, \bar{\vartheta}\psi), \varphi \in \Gamma(A^{(p,q)}), \psi \in \Gamma(A^{(p+1,q)}),$$

$$(d\varphi, \psi) = (\varphi, \delta\psi), \varphi \in \Gamma(A^r), \psi \in \Gamma(A^{r+1}).$$

Proof. (Integration by parts; we only prove the first equation.)
$\chi = \varphi \wedge * \bar{\psi}$ is of type $(n, n - 1)$. Hence, $\partial\chi = 0$ and $d\chi = \bar{\partial}\chi$. Thus,

$$0 = \int_M \partial\chi = \int_M \bar{\partial}\chi = \int_M \bar{\partial}(\varphi \wedge *\bar{\psi})$$

$$= \int_M \bar{\partial}\varphi \wedge *\bar{\psi} + \int_M \varphi \wedge (-1)^{p+q} \bar{\partial}(*\bar{\psi})$$

$$= (\bar{\partial}\varphi, \psi) + (-1)^{p+q} \int_M \varphi \wedge \bar{\partial}(*\bar{\psi}).$$

So

$$(\bar{\partial}\varphi, \psi) = -\int_M \varphi \wedge ** \bar{\partial}(*\bar{\psi})$$

$$= -\int_M \varphi \wedge *(\overline{*\partial(*\psi)})$$

$$= \int_M \varphi \wedge *(\overline{\vartheta\psi}) = (\varphi, \vartheta\psi). \qquad \text{Q.E.D.}$$

If $\varphi = \sum \varphi_{\bar{\beta}} \, dz^{\bar{\beta}}$, then

$$\bar{\partial}\varphi = \frac{1}{2} \sum \left(\frac{\partial \varphi_{\bar{\beta}}}{\partial z^{\bar{\alpha}}} - \frac{\partial \varphi_{\bar{\alpha}}}{\partial z^{\bar{\beta}}} \right) dz^{\bar{\alpha}} \wedge dz^{\bar{\beta}}.$$

Thus the coefficient $(\bar{\partial}\varphi)_{\overline{\alpha\beta}}$ is similar to a "rotation." We claim that ϑ acts like a divergence.

PROPOSITION 2.3.   For $\psi \in \Gamma(A^{(p,q+1)})$,

$$(\vartheta\psi)^{\bar{A}_p\beta_1\cdots\beta_q} = (-1)^{p+1} \sum_{\beta=1}^{n} \left(\frac{\partial}{\partial z^\beta} + \frac{\partial \log g}{\partial z^\beta}\right)\psi^{\bar{A}_p\beta\beta_1\cdots\beta_q}.$$

REMARK.   In Euclidean space div $\psi = \sum (\partial\psi^\beta/\partial z^\beta)$ for a 1-form, $\psi = \sum \psi^\beta \, dx^\beta$.

Proof.   (of the proposition)   By definition

$$(\bar{\partial}\varphi, \psi) = (\varphi, \vartheta\psi).$$

For convenience we omit the index A_p and use the notation

$$\partial_\beta = \frac{\partial}{\partial z^\beta}, \qquad \bar{\partial}_\beta = \frac{\partial}{\partial z^{\bar{\beta}}}.$$

Then

$$(\bar{\partial}\varphi, \psi) = \frac{1}{(q+1)!} \int_M \sum (\bar{\partial}\varphi)_{\bar{\beta}_0\cdots\bar{\beta}_q} \psi^{\bar{\beta}_0\cdots\bar{\beta}_q} 2^n g \, dx^1 \cdots dx^{2n}.$$

Since

$$(\bar{\partial}\varphi)_{\bar{\beta}_0\cdots\bar{\beta}_q} = (-1)^p \sum_{i=0}^{q} \bar{\partial}_{\beta_i} \varphi_{\bar{\beta}_0\cdots\hat{\bar{\beta}}_i\cdots\bar{\beta}_q},$$

where $\hat{\bar{\beta}}_i$ means " omit " $\bar{\beta}_i$. Then

$$(\bar{\partial}\varphi, \psi) = \frac{1}{q!}(-1)^p \int_M \sum \bar{\partial}_\beta \varphi_{\bar{\beta}_i\cdots\bar{\beta}_q} \overline{\psi^{\beta\beta_1\cdots\beta_q}} g 2^n \, dx^1 \cdots dx^{2n}$$

$$= -(-1)^p \frac{1}{q!} \int_M \sum \varphi_{\bar{\beta}_1\cdots\bar{\beta}_q} \overline{\partial_\beta(\psi^{\beta\beta_1\cdots\beta_q}g)} 2^n \, dx^1 \cdots dx^{2n}$$

(integrating by parts).

Thus,

$$(\vartheta\psi)^{\bar{\alpha}_1\cdots\bar{\alpha}_p\beta_1\cdots\beta_q} = (-1)^{p+1} \sum_{\beta=1}^{n} \left(\frac{\partial}{\partial z^\beta} + \frac{1}{g}\frac{\partial g}{\partial z^\beta}\right)\psi^{\bar{\alpha}_1\cdots\bar{\alpha}_p\beta\beta_1\cdots\beta_q}$$

$$= -\sum_{\beta=1}^{n} \left(\frac{\partial}{\partial z^\beta} + \frac{1}{g}\frac{\partial g}{\partial z^\beta}\right)\psi^{\beta\bar{\alpha}_1\cdots\bar{\alpha}_p\beta_1\cdots\beta_q}. \qquad \text{Q.E.D.}$$

Next we make:

DEFINITION 2.3. $\square = \bar{\partial}\vartheta + \vartheta\bar{\partial}$. $\square$ maps $\Gamma(A^{(p,q)})$ into $\Gamma(A^{(p,q)})$ and is called the (complex) *Laplacian.* $\square$ is a partial differential operator and we want to compute its principal part.

PROPOSITION 2.4.   Let $A = (\alpha_1, \cdots, \alpha_p)$, $\beta = (\beta_1, \cdots, \beta_q)$. Then

$$(\square \psi)_{AB} = - \sum_{\lambda, \nu = 1}^{n} g^{\bar{\nu}\lambda} \frac{\partial^2 \varphi_{AB}}{\partial z^\lambda \partial z^{\bar{\nu}}} + \sum_{\lambda, M, N} \left(h_{AB}^{M\bar{N}\lambda} \frac{\partial \varphi_{MN}}{\partial z^\lambda} + h_{AB}^{M\bar{N}\bar{\lambda}} \frac{\partial \varphi_{MN}}{\lambda z^{\bar{\lambda}}} \right)$$

$$+ \sum_{M, N} k_{AB}^{MN} \varphi_{M\bar{N}}.$$

Thus the principal part of $\square$ is

$$- \sum_{\lambda, \nu = 1}^{n} g^{\bar{\nu}\lambda} \frac{\partial^2}{\partial z^\lambda \partial z^{\bar{\nu}}}.$$

REMARK.   If $g^{\bar{\nu}\lambda} = \delta^{\bar{\nu}\lambda}$, then

$$- \sum_{\lambda, \nu} \delta^{\bar{\nu}\lambda} \frac{\partial^2}{\partial z^\lambda \partial z^{\bar{\nu}}} = - \sum_{\nu = 1}^{n} \frac{\partial^2}{\partial z^\nu \partial z^{\bar{\nu}}}$$

$$= - \frac{1}{4} \sum_{k=1}^{2n} \left(\frac{\partial}{\partial x^k} \right)^2.$$

Proof.   (of the proposition)   From Proposition 2.3,

$$(\vartheta \varphi)^{A_1 \beta_1 \cdots \beta_{q-1}} = -(-1)^p \sum \partial_\beta \varphi^{A_p \beta \beta_1 \cdots \beta_{q-1}} + \text{order zero terms.}$$

We note that

$$\partial_\beta \varphi^{\bar{\alpha}_1 \cdots \beta \beta_1 \cdots \beta_{q-1}} = \sum g^{\bar{\alpha}_1 \lambda_1} \cdots g^{\bar{\mu}\beta} \cdot g^{\bar{\mu}_1 \beta_1} \cdots \partial_\beta \varphi_{\lambda_1 \cdots \bar{\mu}\mu_1 \cdots}$$

$$+ \text{terms of order zero.}$$

Thus "lowering indices,"

$$(\vartheta \varphi)_{A_p \bar{\beta}_1 \cdots \bar{\beta}_{q-1}} = (-1)^{p+1} \sum_{\beta, \mu} g^{\bar{\mu}\beta} \partial_\beta \varphi_{A_p \bar{\mu}\bar{\beta}_1 \cdots \bar{\beta}_{q-1}}$$

$$+ \text{order zero terms.}$$

Then

$$(\bar{\partial}\vartheta \varphi)_{A_p \bar{\beta}_1 \cdots \bar{\beta}_q} = -1 \sum_{\beta, \mu} g^{\bar{\mu}\beta} \left\{ \sum_{i=1}^{q} (-1)^{i+1} \bar{\partial}_{\beta_i} \partial_\beta \varphi_{A_p \bar{\mu}\bar{\beta}_i \cdots \hat{\bar{\beta}}_i \cdots \bar{\beta}_q} \right\} \tag{6}$$

$$+ \text{terms of order} \leq 1.$$

On the other hand,

$$(\bar{\partial}\varphi)_{A_p \bar{\beta}_0 \cdots \bar{\beta}_q} = (-1)^p \sum_{i=0}^{q} (-1)^i \bar{\partial}_{\beta_i} \varphi_{A_p \bar{\beta}_0 \cdots \hat{\bar{\beta}}_i \cdots \bar{\beta}_q}.$$

Thus,

$$(\vartheta\bar{\partial}\varphi)_{A_p\bar{\beta}_1\cdots\bar{\beta}_q} = (-1)^{p+1}\sum g^{\bar{\mu}\beta}\,\partial_\beta(\bar{\partial}\varphi)_{A_p\bar{\mu}\bar{\beta}_1\cdots\bar{\beta}_q} \tag{7}$$
$$+ \text{ terms of order} \le 1$$
$$= -\sum_{\beta,\mu} g^{\bar{\mu}\beta}\{\partial_\beta\,\bar{\partial}_\mu\,\varphi_{A_p\bar{\beta}_1\bar{\beta}_2\cdots} - \partial_\beta\,\bar{\partial}_{\beta_1}\,\varphi_{A_p\bar{\mu}\bar{\beta}_2\cdots} + \cdots\}.$$

Many terms cancel when Equations (6) and (7) are added yielding

$$(\Box\varphi)_{A_p\bar{\beta}_1\bar{\beta}_2\cdots\bar{\beta}_q} = -\sum_{\beta,\bar{\mu}} g^{\bar{\mu}\beta}\,\partial_\beta\,\bar{\partial}_\mu\,\varphi_{A_p\bar{\beta}_1\cdots\bar{\beta}_q}$$
$$+ \text{ lower order terms.} \qquad \text{Q.E.D.}$$

Similarly we define $\overline{\Box} = \partial\bar{\vartheta} + \bar{\vartheta}\partial$, $\triangle = d\delta + \delta d$ and prove:

PROPOSITION 2.5. $\overline{\Box} = -\sum g^{\bar{\mu}\beta}\,\partial^2/\partial z^\beta\partial z^{\bar{\mu}} + \text{lower order terms}$
$$\triangle = -\sum 2g^{\bar{\mu}\beta}\,\partial^2/\partial z^\beta\,\partial z^{\bar{\mu}} + \text{lower order terms}.$$

($\triangle$ is the real Laplacian.)

Proof. Left to the reader.

The operators $\overline{\Box}$, $\Box$, and $\triangle$ are second-order partial differential operators. Since $d = \partial + \bar{\partial}$, $\delta = \vartheta + \bar{\vartheta}$, we get

$$\triangle = d\delta + \delta d = (\partial + \bar{\partial})(\vartheta + \bar{\vartheta}) + (\vartheta + \bar{\vartheta})(\partial + \bar{\partial})$$
$$= \Box + \overline{\Box} + \partial\vartheta + \vartheta\partial + \overline{\partial\vartheta + \vartheta\partial}.$$

LEMMA 2.1. $\partial\vartheta + \vartheta\partial$ is a first-order operator.

Proof. $\quad(\vartheta\varphi)_{\alpha_1\cdots\alpha_p\bar{\beta}_2\cdots} = (-1)^{p+1}\sum_{\mu,\beta} g^{\bar{\mu}\beta}\,\partial_\beta\,\varphi_{\alpha_1\cdots\alpha_p\mu\bar{\beta}_1}\cdots$
$$+ \text{ order zero terms.}$$

$$(\partial\vartheta\varphi)_{\alpha_0\cdots\alpha_p\bar{\beta}_1\cdots} = (-1)^{p+1}\sum_{\mu,\beta} g^{\bar{\mu}\beta}\{\partial_{\alpha_0}\,\partial_\beta\,\varphi_{\alpha_1\alpha_2\cdots} - \partial_{\alpha_1}\,\partial_\beta\,\varphi_{\alpha_0\alpha_2\cdots} + \cdots\}$$
$$+ \text{ lower order terms.} \tag{8}$$

$$(\partial\varphi)_{\alpha_0\cdots\alpha_p\bar{\beta}_1\cdots} = \partial_{\alpha_0}\,\varphi_{\alpha_1\alpha_2\cdots} - \partial_{\alpha_1}\varphi_{\alpha_0\alpha_2\cdots} + \cdots$$
$$\tag{9}$$
$$(\vartheta\partial\varphi)_{\alpha_0\cdots\alpha_p\bar{\beta}_1\cdots} = (-1)^p\sum_{\bar{\mu}\,\beta} g^{\bar{\mu}\beta}\,\partial_\beta\{\partial_{\alpha_0}\,\varphi_{\alpha_1\cdots\overline{\mu\beta}_1\cdots} - \partial_{\alpha_1}\varphi_{\alpha_0\alpha_2\cdots\overline{\mu\beta}_1\cdots} + \cdots\}$$
$$+ \text{ lower order terms.}$$

When we sum Equations (8) and (9) the second-order terms cancel. Q.E.D.

COROLLARY. $\Delta = \square + \overline{\square} + $ first-order terms.

For our purposes the following theorem emphasizes the most important fact about $\square$, $\overline{\square}$, and Δ. For the definition and standard facts about elliptic operators we refer the reader to Palais (1965) or Hörmander (1963).

THEOREM 2.2. $\square$, $\overline{\square}$, and Δ are strongly elliptic partial differential operators.

Proof. $g^{\alpha\bar{\beta}}$ is positive definite.   Q.E.D.

3.   Norms for Holomorphic Vector Bundles

The main purpose of this section is to extend the results of Section 2 to vector bundles. Let F be a holomorphic vector bundle over the complex manifold M and $\Gamma(A^{(p,q)}(F))$ the space of C^∞ (p, q)-forms with coefficients in F. Let $\{f_{jkv}^\lambda\}$ be a 1-cocycle defining F on the coordinate covering $\mathcal{U} = \{U_j\}$ of M. Then locally $\varphi \in \Gamma(A^{(p,q)}(F))$ is given by

$$\varphi = (\varphi_j^1(z), \cdots, \varphi_j^m(z)) \text{ on } U_j,$$

where

$$\varphi_j^\lambda(z) = \frac{1}{p!\,q!} \sum \varphi_{j\alpha_1 \cdots \bar{\beta}_1 \cdots}\, dz^{\alpha_1} \wedge \cdots \wedge dz^{\bar{\beta}_1} \wedge \cdots,$$

and on $U_j \cap U_k$,

$$\varphi_j^\lambda(z) = \sum_{v=1}^{m} f_{jkv}^\lambda(z)\varphi_k^v(z).$$

By definition $\bar{\partial}\varphi = (\bar{\partial}\varphi_j^\lambda(z))$ which is well defined since $\bar{\partial}f_{jkv}^\lambda(z) = 0$. Let $2\sum g_{\lambda\bar{v}}\, dz^\lambda\, dz^{\bar{v}}$ be a given Hermitian metric on M. An *Hermitian form on the fibres* of F is defined by specifying on each U_j a positive definite form

$$a_j(\zeta, \zeta) = \sum_{\lambda, v} a_{j\lambda\bar{v}}(z)\zeta_j^\lambda \overline{\zeta_j^v}$$

such that $a_{j\lambda\bar{v}}(z)$ is C^∞ and $a_j(\zeta, \zeta) = a_k(\eta, \eta)$ where $\eta = f_{kj} \cdot \zeta$.

REMARK.   Such forms always exist. Let $\{\rho_j(z)\}$ be a partition of unity subordinate to the locally finite covering $\mathcal{U}$ and set $a(\zeta, \zeta)(z) = \sum_j \rho_j(z) \sum_{\lambda=1}^{m} |\zeta_j^\lambda|^2$.

Let $\varphi, \psi \in \Gamma(A^{(p,q)}(F))$, $\varphi(z) = (\varphi_j^\lambda(z))$, $\psi(z) = (\psi_j^\lambda(z))$.

Then we define a C^∞ function $(\varphi, \psi)(z)$ by

$$(\varphi, \psi)(z) = \sum_{\lambda, \mu=1}^{m} a_{j\lambda\bar{\mu}}(\varphi_j^\lambda, \psi_j^\mu)(z),$$

where $(\varphi_j^\lambda, \psi_j^\mu)(z)$ is the product of (p, q)-forms at z. (See Section 2.) If M is compact we define

$$(\varphi, \psi) = \int_M (\varphi, \psi)(z) \frac{\omega^n}{n!} \qquad (\dim_{\mathbb{C}} M = n)$$

$$= \int_M \sum_{\lambda, \mu = 1}^m a_{j\lambda\bar{\mu}}(z)\varphi_j^\lambda(z) \wedge *\overline{\psi_j^\mu(z)}.$$

We want to define the adjoint ϑ_a of $\bar{\partial}$ (with respect to the metric a). We want to solve

$$(\bar{\partial}\varphi, \psi) = (\varphi, \vartheta_a \psi)$$

for ϑ_a. Let $\varphi \in \Gamma(A^{(p,q)}(F))$, $\psi \in \Gamma(A^{(p,q+1)}(F))$. Then

$$\tau = \sum_{\lambda, \mu = 1}^m a_{j\lambda\bar{\mu}}\varphi_j^\lambda \wedge *\overline{\psi_j^\mu}$$

defines a differential form of type $(n, n - 1)$ and hence $d\tau$ is a $2n$-form. Then

$$0 = \int_M d\tau = \int_M \bar{\partial}\tau = \int_M \sum a_{j\lambda\bar{\nu}}\, \bar{\partial}\varphi_j^\lambda \wedge *\overline{\psi_j^\nu}$$

$$+ (-1)^{p+q} \int_M \sum \varphi_j^\lambda \wedge \overline{\partial(a_{j\lambda\bar{\nu}}*\psi_j^\nu)}.$$

But

$$\int_M \sum \varphi_j^\lambda a_{j\lambda\bar{\mu}} \wedge *\overline{(\vartheta_a \psi)^\mu} = \int_M \sum a_{j\nu\bar{\lambda}}\, \bar{\partial}\varphi_j^\lambda \wedge *\overline{\psi_j^\nu}.$$

Thus,

$$\sum_\mu a_{j\mu\bar{\lambda}}(*\vartheta_a \psi)^\mu = -(-1)^{p+q}\left(\partial \sum_\nu a_{j\nu\bar{\lambda}}(*\psi_j^\nu)\right).$$

Let

$$(a_j^{\bar{\mu}\lambda}) = (a_{j\bar{\mu}\lambda})^{-1}; \qquad \text{that is, } \sum_{\lambda = 1}^m a_j^{\bar{\mu}\lambda} a_{j\lambda\bar{\nu}} = \delta_\nu^\mu.$$

Then

$$(*\vartheta_a \psi)^\mu = -(-1)^{p+q} \sum_\lambda a_j^{\bar{\lambda}\mu}\left(\partial \sum_\nu a_{j\nu\bar{\lambda}}(*\psi_j^\nu)\right),$$

$$(\vartheta_a \psi)^\mu = -\sum_\lambda a_j^{\bar{\lambda}\mu}*\left(\partial \sum_\nu a_{j\nu\bar{\lambda}}(*\psi_j^\nu)\right).$$

Hence:

PROPOSITION 3.1. $(\vartheta_a \psi)^\mu = -\sum_\lambda a_j^{\bar{\lambda}\mu} *\partial(\sum_\nu a_{j\nu\bar{\lambda}}(*\psi_j^\nu)).$

If we expand this out we get

$$(\vartheta_a \psi)_j^\mu = -(*\partial *\psi)_j^\mu - \sum_{\lambda, \nu} a_j^{\bar\lambda \mu} *(\partial a_{j\nu\lambda} \wedge *\psi_j^\nu)$$

$$= (\vartheta \psi)_j^\mu + \text{terms of order zero.}$$

The reader can then easily verify the following: Let

$$\Box_a = \bar\partial \vartheta_a + \vartheta_a \bar\partial,$$

then

$$\Box_a = \Box + \text{terms of order} \leq 1.$$

Hence:

THEOREM 3.1. $\Box_a = -\sum g^{\beta\alpha} \partial^2/\partial z^\alpha \partial z^\beta + \text{lower order terms and hence, } \Box_a$ is a strongly elliptic second-order operator.

4.   Applications of Results on Elliptic Operators

For the results about elliptic operators on manifolds that we need Palais (1965) is a good source. First we fix some notation: Let $\mathscr{L}^q = \Gamma(A^{(p,\,q)}(F))$ where F is some holomorphic vector bundle. We drop the subscript a and let $\vartheta = \vartheta_a$, $\Box = \Box_a$.

PROPOSITION 4.1. $\Box$ is self-adjoint, that is,

$$(\Box \varphi, \psi) = (\varphi, \Box \psi).$$

Proof. $(\Box \varphi, \psi) = ((\bar\partial \vartheta + \vartheta \bar\partial)\varphi, \psi)$

$$= (\vartheta\varphi, \vartheta\psi) + (\bar\partial\varphi, \bar\partial\psi)$$

$$= (\varphi, (\bar\partial \vartheta + \vartheta \bar\partial)\psi)$$

$$= (\varphi, \Box \psi). \qquad \text{Q.E.D.}$$

The following is the fundamental result about elliptic operators on *compact* manifolds: Let

$$\mathscr{H}^q = \{\varphi \mid \varphi \in \mathscr{L}^q, \Box \varphi = 0\}.$$

THEOREM.   (a) dim $\mathscr{H}^q < +\infty$.   (b) $\mathscr{L}^q = \mathscr{H}^q \oplus \Box \mathscr{L}^p$, where $\oplus$ means orthogonal direct sum; so every $\varphi \in \mathscr{L}^q$ has a unique representation,

$$\varphi = \eta + \zeta, \qquad \eta \in \mathscr{H}^q, \qquad \zeta = \Box \psi.$$

PROPOSITION 4.2. $\square\varphi = 0$ if and only if $\bar\partial\varphi = \vartheta\varphi = 0$.

Proof. $(\square\varphi, \varphi) = (\bar\partial\varphi, \bar\partial\varphi) + (\vartheta\varphi, \vartheta\varphi) = \|\bar\partial\varphi\|^2 + \|\vartheta\varphi\|^2$.   Q.E.D.

PROPOSITION 4.3. $\mathscr{L}^q = \mathscr{H}^q \oplus \bar\partial\mathscr{L}^{q-1} \oplus \vartheta\mathscr{L}^{q+1}$.

Proof.   Take $\eta \in \mathscr{H}^q$, $\bar\partial\varphi \in \bar\partial\mathscr{L}^{q-1}$, $\vartheta\psi \in \vartheta\mathscr{L}^{q+1}$. Then

$$(\eta, \bar\partial\varphi) = (\vartheta\eta, \varphi) = 0,$$
$$(\eta, \vartheta\psi) = (\bar\partial\eta, \psi) = 0,$$
$$(\bar\partial\varphi, \vartheta\psi) = (\overline{\partial\partial}\varphi, \psi) = 0.$$

Thus, $\mathscr{H}^q$, $\bar\partial\mathscr{L}^{q-1}$, and $\vartheta\mathscr{L}^{q+1}$ are orthogonal and

$$\begin{aligned}
\mathscr{L}^q = \mathscr{H}^q \oplus \square\mathscr{L}^q &= \mathscr{H}^q \oplus \bar\partial\vartheta\mathscr{L}^q \oplus \vartheta\bar\partial\mathscr{L}^q \\
&\subseteq \mathscr{H}^q \oplus \bar\partial\mathscr{L}^q \oplus \vartheta\mathscr{L}^q \\
&\subseteq \mathscr{L}^q. \qquad \text{Q.E.D.}
\end{aligned}$$

We next have the important theorem relating cohomology groups and harmonic forms.

THEOREM 4.1.   [Kodaira (1953)]   Let F be a holomorphic vector bundle on a compact complex manifold. If $\Omega^p(F)$ is the sheaf of germs of holomorphic p-forms with values in F, then

$$H^q(M, \Omega^p(F)) \cong \mathscr{H}^q.$$

Proof.   Recall $H^p(M, \Omega^p(F)) \cong \Gamma(\bar\partial A^{(p,q-1)}(F))/\bar\partial\Gamma(A^{(p,q-1)}(F))$ and

$$\Gamma(\bar\partial A^{(p,q-1)}(F)) = \{\varphi \mid \varphi \in \Gamma(A^{(p,q)}(F)), \bar\partial\varphi = 0\}.$$

Let $Z_{\bar\partial}(\mathscr{L}^q) = \{\varphi \mid \varphi \in \mathscr{L}^q, \bar\partial\varphi = 0\}$. Then

$$H^q(M, \Omega^p(F)) \cong \frac{Z_{\bar\partial}(\mathscr{L}^q)}{\bar\partial\mathscr{L}^{q-1}}.$$

We claim

$$Z_{\bar\partial}(\mathscr{L}^q) = \mathscr{H}^q \oplus \bar\partial\mathscr{L}^{q-1}. \tag{1}$$

The inclusion

$$\mathscr{H}^q \oplus \bar\partial\mathscr{L}^{q-1} \subseteq Z_{\bar\partial}(\mathscr{L}^q)$$

is obvious. Let $\varphi \in Z_{\bar\partial}(\mathscr{L}^q)$, $\varphi = \eta + \bar\partial\psi + \vartheta\sigma$. Then $\bar\partial\varphi = 0$ so

$$\bar\partial\vartheta\sigma = 0$$
$$(\bar\partial\vartheta\sigma, \sigma) = 0$$
$$(\vartheta\sigma, \sigma) = 0$$

and

$$\vartheta\sigma = 0.$$

Thus Equation (1) is true and implies the theorem.   Q.E.D.

COROLLARY. $\dim H^q(M, \Omega^p(F)) < +\infty$.

EXAMPLE.   Let $M^n = \mathbb{C}^n/G$ be a complex torus. If $z = (z^1, \cdots, z^n)$ are coordinates on $\mathbb{C}^n$, then $\omega = i\sum_{\alpha=1}^n dz^\alpha \wedge dz^{\bar\alpha}$ defines a 2-form on M^n associated to the metric on M^n lifted from the Euclidean metric of $\mathbb{C}^n$. If $z^\alpha = x^{2\alpha-1} + i\,x^{2\alpha}$ then $\square = -\frac{1}{4}\sum_{k=1}^{2n}(\partial/\partial x^k)^2$. Let

$$\varphi = \frac{1}{p!\,q!}\sum \varphi_{\alpha_1\cdots\alpha_p\bar\beta_1\cdots\bar\beta_q}\, dz^{\alpha_1} \wedge \cdots \wedge dz^{\bar\beta_q}.$$

Then $\square\varphi = 0$ if and only if $\sum_{k=1}^{2n}(\partial/\partial x^k)^2\,\varphi_{A\bar{B}} = 0$. Such solutions $\varphi_{A\bar B}$ are necessarily constant so

$$\varphi = \frac{1}{p!\,q!}\sum c_{\alpha_1\cdots\alpha_p\bar\beta_1\cdots\bar\beta_q}\, dz^{\alpha_1} \wedge \cdots \wedge dz^{\bar\beta_q}.$$

Thus $\dim_{\mathbb{C}} \mathscr{H}^{p,q} = \binom{n}{p}\binom{n}{q}$ so

$$\dim_{\mathbb{C}} H^q(M^n, \Omega^p) = \binom{n}{p}\binom{n}{q}.$$

THEOREM 4.2.   [Serre duality; see Serre (1955)]   Let F, M be as in Theorem 4.1. Let F^* be the dual bundle of F. Then

$$H^q(M^n, \Omega^p(F)) \cong H^{n-q}(M^n, \Omega^{n-p}(F)).$$

Proof. Let $\omega = i\sum g_{\alpha\bar\beta}\, dz^\alpha \wedge dz^{\bar\beta}$ be the form associated to a metric on M. If F is represented by $\{f^\lambda_{jk\nu}\}$ on $\mathscr{U} = \{U_j\}$ (that is, $\zeta^\lambda_j = \sum_{\nu=1}^m f^\lambda_{jk\nu}\,\zeta^\nu_k$ on $U_j \cap U_k$) then F^* is represented by $\{f^\nu_{kj\lambda}\}$ (that is, $\eta_{j\lambda} = \sum_{\nu=1}^m f^\nu_{kj}\,\eta_{k\lambda}$ on $U_j \cap U_k$) and $\sum_\mu f^\mu_{kj\lambda} \cdot f^\nu_{jk\mu} = \delta^\nu_\lambda$. If $a_j(\zeta, \zeta) = \zeta_j\,a_j\,\bar\zeta^t_j$ (matrix notation; $A^t =$ transpose of A) is an Hermitian form on F it is easy to see that a transforms as follows:

$$a_k = f^t_{jk}\,a_j \cdot \bar f_{jk}. \tag{2}$$

We define a conjugate linear map $\rho: F \to F^*$ by $\rho(\zeta_j) = \sum a_{j\lambda\bar\nu}\bar\zeta^\nu_j$, that is, $\rho(\zeta_j) \in F^*$ acts on η_j by $\rho(\zeta_j)(\eta_j) = \sum a_{j\lambda\bar\nu}\,\eta^\lambda_j\bar\zeta^\nu_j$. From Equation (2) we see this is well defined. Using ρ we introduce a metric on the fibres of F^* by

$$b(\eta, \eta) = a(\rho^{-1}\eta, \rho^{-1}\eta).$$

It is easy to check that the matrix of $b(\eta, \eta)$ is

$$b_j = a_j^{-1} = (a_j^{\bar{\nu}\lambda}) \text{ on } U_j.$$

Let

$$\mathcal{H}^{p,q}(F) = \{\varphi \mid \varphi \in \Gamma(A^{p,q}(F)), \Box\varphi = 0\}$$

$$\mathcal{H}^{n-p,n-q}(F^*) = \{\varphi \mid \varphi \in \Gamma(A^{n-p,n-q}(F^*)), \Box\varphi = 0\}.$$

If F were trivial we could define a map,

$$\Gamma(A^{p,q}) \xrightarrow{\quad-\quad} \Gamma(A^{q,p}) \xrightarrow{\quad*\quad} \Gamma(A^{n-p,n-q})$$

$$\varphi \longrightarrow \bar{\varphi} \longrightarrow *\bar{\varphi} = \#\varphi.$$

In general we define $\#\varphi$ by

$$(\#\varphi)_{j\lambda} = \sum a_{j\lambda\bar{\nu}}(\overline{*\varphi_j^\nu}) = \rho \circ *.$$

LEMMA 4.1. $\# : \Gamma(A^{p,q}(F)) \to \Gamma(A^{n-p,n-q}(F^*))$ satisfies

$$\# \circ \# = (-1)^{p+q} id,$$

where id = identity.

Proof. $(\#(\#\varphi))_j^\lambda = \sum a_j^{\bar{\mu}\lambda} \overline{a_{j\mu\bar{\nu}}(\overline{**\varphi})_j^\nu}$

$$= (-1)^{p+q}\varphi_j^\lambda.   \text{Q.E.D.}$$

Recall

$$(\vartheta\varphi)_j^\lambda = -\sum a_j^{\bar{\mu}\lambda} \cdot \partial(a_{j\nu\bar{\mu}} * \varphi_j^\nu);$$

and since a_j is nonsingular

$$(\vartheta\varphi)_j^\nu = 0 \quad \text{if and only if} \quad \partial\left(\sum_\nu a_{j\nu\bar{\mu}} * \varphi_j^\nu\right) = 0.$$

Conjugate this to get

$$\bar{\partial}\left(\sum_\nu a_{j\mu\bar{\nu}} * \overline{\varphi_j^\nu}\right) = 0,$$

that is

$$\bar{\partial}(\#\varphi)_{j\mu} = 0.$$

Thus

$$\mathcal{H}^{p,q}(F) = \{\varphi \mid \bar{\partial}\varphi_j^\lambda = 0, \bar{\partial}(\#\varphi)_{j\lambda} = 0\},$$

and

$$\mathcal{H}^{n-p,\,n-q}(F^*) = \{\psi \mid \bar{\partial}\psi_{j\lambda} = 0,\ \bar{\partial}(\#\psi)_j^\lambda = 0\}.$$

So

$$\# : \mathcal{H}^{p,\,q}(F) \longrightarrow \mathcal{H}^{n-p,\,n-q}(F)$$

and

$$\# : \mathcal{H}^{n-p,\,n-q}(F) \longrightarrow \mathcal{H}^{p,\,q}(F).$$

Lemma 4.1 implies that $\#$ is a (conjugate linear) isomorphism.   Q.E.D.

EXAMPLE.   Let M be a compact Riemann surface. Then

$$H^1(M, \mathcal{O}) = H^1(M, \Omega^0) \cong H^0(M, \Omega^1)$$

and

$$H^0(M, \Omega^1) \text{ is the space of } holomorphic \ differentials$$

on M. Thus,

$$\dim H^1(M, \mathcal{O}) = \dim H^0(M, \Omega^1) = \text{genus of } M.$$

Further

$$H^1(M, \Theta) = H^1(M, \mathcal{O}(T)) \cong H^0(M, \Omega^1(T^*))$$
$$= H^0(M, \mathcal{O}(T^* \otimes T^*))$$

which is the space of *holomorphic quadratic differentials* on M. Thus,

$$\dim H^1(M, \Theta) = \begin{cases} 0, & \text{if genus } (M) = 0 \\ 1, & \text{if genus } (M) = 1 \\ 3g - 3, & \text{if genus } (M) \geq 2. \end{cases}$$

[For example, see Teichmüller (1940).]

5.   Covariant Differentiation on Kähler Manifolds

In this section we want to exhibit some of the special facts that a Kählerian structure imposes on the Hermitian geometry of M, a complex manifold. For instance, $\Box = \overline{\Box} = \tfrac{1}{2}\Delta$ holds on a Kähler manifold. First we must review the idea of a covariant derivative.

Suppose $\sum_\alpha \xi_j^\alpha(z)(\partial/\partial z_j^\alpha) = \sum_\alpha \xi_k^\alpha(z)(\partial/\partial z_k^\alpha)$ is a given C^∞ section of T. If $z_i \in U_i \cap U_j \cap U_k \neq \phi$, then in general,

$$\sum_\alpha \frac{\partial \xi_j^\alpha}{\partial z_i^\lambda}\left(\frac{\partial}{\partial z_j^\alpha}\right) \neq \sum_\alpha \frac{\partial \xi_k^\alpha}{\partial z_i^\lambda}\left(\frac{\partial}{\partial z_k^\alpha}\right)$$

because

$$\xi_j^\alpha = \sum_\beta \frac{\partial z_j^\alpha}{\partial z_k^\beta} \xi_k^\beta$$

and thus,

$$\frac{\partial \xi_j^\alpha}{\partial z_i^\lambda} = \sum \frac{\partial z_j^\alpha}{\partial z_k^\beta} \frac{\partial \xi_k^\beta}{\partial z_i^\lambda} + \sum \frac{\partial^2 z_j^\alpha}{\partial z_i^\lambda \, \partial z_k^\beta} \xi_k^\beta .$$

We would like to define a "correction" term $\Gamma_{j\lambda\beta}^\alpha(z)$ such that

$$\sum_\alpha \nabla_\lambda \xi_j^\alpha \left(\frac{\partial}{\partial z_j^\alpha} \right) = \sum \nabla_\lambda \xi_k^\beta \left(\frac{\partial}{\partial z_k^\beta} \right),$$

where

$$\nabla_\lambda \xi_j^\alpha = \frac{\partial \xi_j^\alpha}{\partial z_i^\lambda} + \sum_\beta \Gamma_{j\lambda\beta}^\alpha \xi_j^\beta .$$

We temporarily fix the following notation: C^∞ sections in $\mathscr{D}(T)$, $\mathscr{D}(T^*)$ $\mathscr{D}(\overline{T})$, $\mathscr{D}(\overline{T}^*)$ will be written

$$\sum \xi_j^\alpha \left(\frac{\partial}{\partial z_j^\alpha} \right), \quad \sum \varphi_{j\alpha}\, dz_j^\alpha, \quad \sum \eta_j^{\bar\alpha} \left(\frac{\partial}{\partial z_j^\alpha} \right), \quad \sum \psi_{j\alpha}\, dz_j^{\bar\alpha},$$

respectively. We suppose we have fixed an Hermitian metric $ds^2 = \sum g_{j\alpha\bar\beta}$ $dz_j^\alpha \wedge dz_j^{\bar\beta}$ on M. Later we will assume that this metric is Kähler. Let U be a coordinate patch with coordinates $(z^1, \cdots, z^n)$. Let $\partial_\lambda - \partial/\partial z^\lambda$, $\bar\partial_\lambda - \partial/\partial \bar z^\lambda$. The $\eta_j^{\bar\alpha}$ transform as follows:

$$\eta_j^{\bar\alpha}(z) = \sum_\beta \left(\overline{\frac{\partial z_j^\alpha}{\partial z_k^\beta}} \right) \eta_k^{\bar\beta}(z).$$

Since

$$\partial_\lambda \left(\overline{\frac{\partial z_j^\alpha}{\partial z_k^\beta}} \right) = 0,$$

$$\partial_\lambda \eta_j^{\bar\alpha}(z) = \sum_\beta \left(\overline{\frac{\partial z_j^\alpha}{\partial z_k^\beta}} \right) \partial_\lambda \eta_k^{\bar\beta}(z).$$

Similarly for $\psi_{j\alpha}$. We define

$$\nabla_\lambda \eta_j^{\bar\alpha} = \partial_\lambda \eta_j^{\bar\alpha}, \, \nabla_\lambda \psi_{j\bar\alpha} = \partial_\lambda \psi_{j\bar\alpha} . \tag{1}$$

Let $(g_j^{\bar\beta\alpha}) = (g_{j\alpha\bar\beta})^{-1}$. Then

$$\rho(\xi) = \sum_{\gamma=1}^n g_{j\gamma\bar\beta}\, \xi_j^\gamma\, dz^{\bar\beta} \in \mathscr{D}(\overline{T}^*)$$

and we define

$$\nabla_\lambda \xi = \rho^{-1}\, \nabla_\lambda \rho(\xi),$$

that is,

$$\nabla_\lambda \xi_j^\alpha = \sum_\beta g_j^{\bar\beta\alpha}\, \partial_\lambda \Big(\sum g_{j\gamma\bar\beta}\, \xi_j^\gamma\Big).$$

Thus,

$$\nabla_\lambda \xi_j^\alpha = \partial_\lambda \xi_j^\alpha + \sum_{\gamma,\beta} (g_j^{\bar\beta\alpha}\, \partial_\lambda g_{j\gamma\bar\beta})\xi_j^\gamma$$

and

$$\Gamma^\alpha_{\lambda\gamma} = \sum_\beta g_j^{\bar\beta\alpha}\, \partial_\lambda g_{\gamma\bar\beta}$$

$$= (\partial_\lambda G)\cdot G^{-1},$$

where $G = (g_{\alpha\bar\beta})$.
  Similarly,

$$\nabla_\lambda \varphi_{j\alpha} = \sum_{\gamma,\beta} g_{j\alpha\bar\beta}\, \bar\partial_\lambda (g_j^{\bar\beta\gamma}\varphi_{j\gamma})$$

$$= \partial_\lambda \varphi_{j\alpha} - \sum_{\gamma=1}^{n} \Gamma^\gamma_{\lambda\alpha}\, \varphi_{j\gamma},$$

since

$$\sum_\beta g_{\alpha\bar\beta}\, g^{\bar\beta\gamma} = \delta^\gamma_\alpha,$$

and

$$0 = \sum_\beta \partial_\lambda g_{\alpha\bar\beta}\, g^{\bar\beta\gamma} + \sum_\beta g_{\alpha\bar\beta}\, \partial_\lambda g^{\bar\beta\gamma}.$$

We also notice that if $(\tilde z^1, \cdots, \tilde z^n)$ are different coordinates on U,

$$\left(\frac{\partial}{\partial \tilde z^\lambda}\right) = \sum_\mu \frac{\partial z^\mu}{\partial \tilde z^\lambda}\left(\frac{\partial}{\partial z^\mu}\right),$$

so

$$\tilde\nabla_\lambda = \sum_{\mu=1}^{n} \frac{\partial z^\mu}{\partial \tilde z^\lambda}\, \nabla_\mu.$$

Suppose now that $z_j = z$. Then we see that

$$\nabla_\lambda \xi^\alpha\, dz^\lambda \otimes \left(\frac{\partial}{\partial z^\alpha}\right) \in \mathscr{D}(T^* \otimes T)$$

and

$$\nabla_\lambda \varphi_\alpha\, dz^\lambda \otimes dz^\alpha \in \mathscr{D}(T^* \otimes T^*),$$

that is,

$$\nabla_{j\lambda}\xi_j^{\alpha} = \sum_{\mu,\beta} \frac{\partial z_k^{\mu}}{\partial z_j^{\lambda}} \frac{\partial z_j^{\alpha}}{\partial z_k^{\beta}} \nabla_{k\mu}\xi_k^{\beta}.$$

This remark is used in differential geometry as motivation for defining $\nabla : \mathscr{D}(E) \to \mathscr{D}(T^* \otimes E)$, where E is a C^{∞} vector bundle over M a differentiable manifold. We also define $\bar{\nabla}$,

$$\bar{\nabla}_{\lambda}\xi^{\alpha} = \bar{\partial}_{\lambda}\xi^{\alpha}, \ \bar{\nabla}_{\lambda}\varphi_{\alpha} = \bar{\partial}_{\lambda}\varphi_{\alpha},$$

$$\bar{\nabla}_{\lambda}\eta^{\bar{\alpha}} = \bar{\partial}_{\lambda}\eta^{\bar{\alpha}} + \sum_{\beta} \overline{\Gamma^{\alpha}_{\lambda\beta}}\, \eta^{\bar{\beta}},$$

$$\bar{\nabla}_{\lambda}\psi_{\bar{\alpha}} = \bar{\partial}_{\lambda}\psi_{\bar{\alpha}} - \sum_{\beta} \overline{\Gamma^{\beta}_{\lambda\alpha}}\, \psi_{\bar{\beta}}.$$

In fact, we could define $\nabla_{\lambda}\varphi$ analogously for any tensor field

$$\varphi \in \mathscr{D}(T \otimes \cdots \otimes \bar{T} \otimes \cdots T^* \otimes \cdots \bar{T}^*),$$

by raising or lowering indices until $\rho(\varphi) \in \mathscr{D}(\bar{T} \otimes \cdots \bar{T}^*)$ taking ∂_{λ} and then ρ^{-1}. We will not write out the result in local coordinates here.

PROPOSITION 5.1. $\nabla_{\lambda} g_{\alpha\bar{\beta}} = 0, \ \nabla_{\lambda} g^{\alpha\bar{\beta}} = 0$.

Proof. $\quad \nabla_{\lambda} g_{\alpha\bar{\beta}} = \sum g_{\alpha\bar{\sigma}} \partial_{\lambda}(g^{\bar{\sigma}\gamma} g_{\gamma\bar{\beta}})$

$$= \sum g_{\alpha\bar{\sigma}} \partial_{\lambda}(\delta^{\bar{\sigma}}_{\bar{\beta}})$$

$$= 0. \hspace{6cm} \text{Q.E.D}$$

THEOREM 5.1. $\omega = i \sum g_{\alpha\bar{\beta}}\, dz^{\alpha} \wedge dz^{\bar{\beta}}$ is Kähler if and only if

$$\Gamma^{\alpha}_{\gamma\beta} = \Gamma^{\alpha}_{\beta\gamma}.$$

Proof. ω is Kähler if and only if $d\omega = 0$ and

$$d\omega = i \sum \frac{\partial g_{\alpha\bar{\beta}}}{\partial z^{\lambda}}\, dz^{\lambda} \wedge dz^{\alpha} \wedge dz^{\bar{\beta}}$$

$$+ i \sum \frac{\partial g_{\alpha\bar{\beta}}}{\partial z^{\bar{\lambda}}}\, dz^{\bar{\lambda}} \wedge dz^{\alpha} \wedge dz^{\bar{\beta}}.$$

Now $\Gamma^{\alpha}_{\lambda\beta} = \sum_{\sigma} g^{\bar{\sigma}\alpha} \partial_{\lambda}(g_{\beta\bar{\sigma}})$ implies $\partial_{\lambda} g_{\beta\bar{\sigma}} = \sum_{\alpha} g_{\alpha\bar{\sigma}} \Gamma^{\alpha}_{\lambda\beta}$, and so on. Therefore,

$$d\omega = \frac{i}{2} \sum g_{\tau\bar{\beta}}(\Gamma^{\tau}_{\alpha\lambda} - \Gamma^{\tau}_{\lambda\alpha})\, dz^{\lambda} \wedge dz^{\alpha} \wedge dz^{\bar{\beta}}$$

$$+ \frac{i}{2} \sum g_{\alpha\bar{\tau}}(\overline{\Gamma^{\tau}_{\lambda\beta}} - \overline{\Gamma^{\tau}_{\beta\lambda}})\, dz^{\bar{\lambda}} \wedge dz^{\alpha} \wedge dz^{\bar{\beta}}.$$

Thus, $d\omega = 0$ if and only if $\Gamma^{\alpha}_{\gamma\beta} = \Gamma^{\alpha}_{\beta\gamma}$.   Q.E.D.

PROPOSITION 5.2. Let $\varphi = 1/p!q! \sum \varphi_{\alpha_1 \cdots \alpha_p \bar{\beta}_1 \cdots \bar{\beta}_q} \, dz^{\alpha_1} \wedge \cdots \wedge dz^{\bar{\beta}_q}$. If M is a Kähler manifold, then

$$\partial\varphi = \frac{1}{p!q!} \sum \nabla_{\alpha} \varphi_{\alpha_1 \cdots \alpha_p \cdots \bar{\beta}_q} \, dz^{\alpha} \wedge dz^{\alpha_1} \cdots dz^{\bar{\beta}_q}.$$

Proof. We prove the case $q = 0$.

$$\frac{1}{p!} \sum \nabla_{\alpha} \varphi_{\alpha_1 \cdots \alpha_p} \, dz^{\alpha} \wedge dz^{\alpha_1} \cdots dz^{\alpha_p}$$

$$= \frac{1}{p!} \sum \Bigg\{ \partial_{\alpha} \varphi_{\alpha_1 \cdots \alpha_p} - \sum_{\sigma} \Gamma^{\sigma}_{\alpha\alpha_1} \varphi_{\sigma\alpha_2 \cdots \alpha_p}$$

$$- \sum_{\sigma} \Gamma^{\sigma}_{\alpha\alpha_2} \varphi_{\alpha_1 \sigma\alpha_3 \cdots \alpha_p} \cdots \Bigg\} \, dz^{\alpha} \wedge dz^{\alpha_1} \wedge \cdots \wedge dz^{\alpha_p}.$$

Since $\Gamma^{\sigma}_{\alpha\alpha_j} = \Gamma^{\sigma}_{\alpha_j\alpha}$ and $dz^{\alpha} \wedge \cdots \wedge dz^{\alpha_p}$ is skew-symmetric all terms sum to zero except the first which sums to $\partial\varphi$.   Q.E.D.

Similarly:

PROPOSITION 5.3. In the Kähler case,

$$\bar{\partial}\varphi = \sum \bar{\nabla}_{\alpha} \varphi_{\alpha_1 \cdots \alpha_p \bar{\beta}_1 \cdots \bar{\beta}_q} \, dz^{\alpha} \wedge dz^{\alpha_1} \wedge \cdots \wedge dz^{\bar{\beta}_q}.$$

THEOREM 5.2. In the Kähler case

$$(\vartheta\varphi)_{\alpha_1 \cdots \bar{\beta}_q} = -(-1)^p \sum_{\beta,\gamma} g^{\beta\gamma} \nabla_{\gamma} \varphi_{\alpha_1 \cdots \alpha_p \bar{\beta}\bar{\beta}_1 \cdots \bar{\beta}_q}.$$

Proof. By Proposition 2.3

$$-(-1)^p (\vartheta\varphi)^{\bar{\alpha}_1 \cdots \bar{\alpha}_p \bar{\beta}_1 \cdots \bar{\beta}_q} = \sum_{\beta} \left(\partial_{\beta} + \frac{1}{g} \partial_{\beta} g \right) \varphi^{\bar{\alpha}_1 \cdots \bar{\alpha}_p \beta\bar{\beta}_1 \cdots \bar{\beta}_q}.$$

By definition

$$\sum_{\beta} \nabla_{\beta} \varphi^{\bar{A}_p \beta\bar{\beta}_1 \cdots \bar{\beta}_q} = \sum_{\beta} \partial_{\beta} \varphi^{\bar{A}_p \beta\bar{\beta}_1 \cdots \bar{\beta}_q} + \sum_{\gamma,\beta} \Gamma^{\beta}_{\beta\gamma} \varphi^{\bar{A}_p \bar{\beta}_1 \cdots \bar{\beta}_p}$$

$$+ \sum_{\gamma,\beta} \Gamma^{\beta_1}_{\beta\gamma} \varphi^{\bar{A}_p \beta\gamma\bar{\beta}_2 \cdots \bar{\beta}_q} + \cdots$$

$$= \sum_{\beta} \partial_{\beta} \varphi^{\bar{A}_p \beta\bar{\beta}_1 \cdots \bar{\beta}_q} + \sum_{\gamma,\beta} \Gamma^{\beta}_{\beta\gamma} \varphi^{\bar{A}_p \bar{\beta}_1 \cdots \bar{\beta}_p}.$$

This is true because $\Gamma^{\beta_k}_{\beta\gamma} = \Gamma^{\beta_k}_{\gamma\beta}$ and $\varphi^{\bar{A}_p\beta\beta_1\cdots\gamma\cdots\beta_q}$ is skew-symmetric in $(\beta\beta_1\cdots\gamma\cdots\beta_q)$. We claim

$$\sum \Gamma^\beta_{\beta\gamma} = \frac{1}{g}\,\partial_\gamma g, \qquad \text{where } g = \det(g_{\alpha\bar\beta}). \tag{2}$$

If we know (2), then

$$\sum_\beta \nabla_\beta \varphi^{\bar{A}_p\beta\beta_1\cdots\beta_q} = \sum_\beta \left(\partial_\beta + \frac{1}{g}\,\partial_\beta g\right)\varphi^{\bar{A}_p\beta\cdots\beta_q} \tag{3}$$

$$= -(-1)^{p+1}(\vartheta\varphi)^{\bar{A}_p\beta\cdots\beta_q}$$

by Proposition 2.3. We remark that

$$\nabla_\lambda \zeta^{\bar\alpha\beta} = (\nabla_\lambda \zeta)^{\bar\alpha\beta},$$

since

$$\zeta^{\bar\alpha\beta} = \sum g^{\bar\alpha\gamma}g^{\bar\delta\beta}\zeta_{\gamma\bar\delta}$$

and

$$\nabla_\lambda g^{\alpha\bar\beta} = 0;$$

thus

$$\nabla_\lambda \zeta^{\bar\alpha\beta} = \sum g^{\bar\alpha\gamma}g^{\bar\delta\beta}\nabla_\lambda \zeta_{\gamma\bar\delta}.$$

Similarly we may prove that (3) implies

$$-(-1)^p \sum_\beta \nabla_\beta \varphi^\beta_{A_p\bar{B}_q} = (\vartheta\varphi)_{A_p\bar{B}_q},$$

so

$$-(-1)^p \sum_{\beta,\gamma} g^{\bar\beta\gamma}\nabla_\gamma \varphi_{A_p\bar\beta\bar\beta_1\cdots\bar\beta_q} = (\vartheta\varphi)_{A_p\bar{B}_q}.$$

Thus we prove (2). Recall

$$\Gamma^\alpha_{\lambda\beta} = \sum_\sigma g^{\bar\sigma\alpha}\,\partial_\lambda(g_{\beta\bar\sigma}).$$

Let $A_{\lambda\bar\nu}$ be the cofactor of $g_{\lambda\bar\nu}$. Then

$$g^{\bar\nu\lambda} = \frac{A_{\lambda\bar\nu}}{g}.$$

But

$$\frac{\partial g}{\partial g_{\lambda\bar\nu}} = A_{\lambda\bar\nu},$$

so

$$\frac{\partial g}{\partial g_{\lambda\bar\nu}} = g\,g^{\bar\nu\lambda}. \tag{4}$$

Using (4) and the fact that M is Kähler

$$\partial_\gamma g = \sum_{\lambda,\bar{\nu}} \frac{\partial g}{\partial g_{\lambda\bar{\nu}}} \frac{\partial g_{\lambda\bar{\nu}}}{\partial z^\gamma} = \sum_{\lambda,\bar{\nu}} g g^{\bar{\nu}\lambda} \partial_\gamma(g_{\lambda\bar{\nu}})$$

$$= g \sum_\lambda \Gamma^\lambda_{\gamma\lambda}$$

$$= g \sum_\lambda \Gamma^\lambda_{\lambda\gamma}. \qquad \text{Q.E.D.}$$

We now introduce an operator Λ mapping $\Gamma(A^{p,q})$ into $\Gamma(A^{p-1,q-1})$. Let

$$\varphi = \frac{1}{p!\,q!} \sum \varphi_{\alpha_1 \cdots \alpha_p \bar{\beta}_1 \cdots \bar{\beta}_q}\, dz^{\alpha_1} \wedge \cdots \wedge dz^{\bar{\beta}_q}.$$

Then

$$\Lambda\varphi = \frac{1}{(p-1)!(q-1)!} \sum i g^{\bar{\beta}\alpha} \varphi_{\alpha\bar{\beta}\alpha_2 \cdots \alpha_p \bar{\beta}_2 \cdots \bar{\beta}_q}\, dz^{\alpha_2} \wedge \cdots \wedge dz^{\bar{\beta}_q},$$

where $i = \sqrt{-1}$; that is,

$$(\Lambda\varphi)_{\alpha_2 \cdots \alpha_p \bar{\beta}_2 \cdots \bar{\beta}_q} = \sum_{\alpha,\beta} i g^{\bar{\beta}\alpha} \varphi_{\alpha\bar{\beta}\alpha_2 \cdots \alpha_p \bar{\beta}_2 \cdots \bar{\beta}_q}$$

$$= (-1)^{p-1} \sum_{\alpha,\beta} i g^{\bar{\beta}\alpha} \varphi_{\alpha\alpha_2 \cdots \alpha_p \bar{\beta}\bar{\beta}_2 \cdots \bar{\beta}_q}.$$

If $p - 1 < 0$ or $q - 1 < 0$ then we set $\Gamma(A^{p-1,q-1}) = 0$.

LEMMA 5.1. Λ is a real operator; that is,

$$\overline{\Lambda\varphi} = \Lambda\bar{\varphi}.$$

Proof. Obvious.

PROPOSITION 5.4. $\partial\Lambda - \Lambda\partial = i\vartheta$, $\bar{\partial}\Lambda - \Lambda\bar{\partial} = -i\vartheta$.

Proof. The first implies the second. We shall prove the first, by Proposition 5.2,

$$(\partial\Lambda\varphi)_{\alpha_1 \cdots \alpha_p \bar{B}} = \nabla_{\alpha_1}(\Lambda\varphi)_{\alpha_2 \cdots \alpha_p \bar{B}} - \nabla_{\alpha_2}(\Lambda\varphi)_{\alpha_1\alpha_3 \cdots \alpha_p \bar{B}} + \cdots,$$

where $\bar{B} = \bar{\beta}_2 \cdots \bar{\beta}_q$. Thus,

$$(\partial\Lambda\varphi)_{\alpha_1 \cdots \alpha_p \bar{B}} = i(-1)^{p-1} \sum_{\alpha,\beta} \{\nabla_{\alpha_1} g^{\bar{\beta}\alpha}\varphi_{\alpha\alpha_2 \cdots \bar{B}} - \nabla_{\alpha_2} g^{\bar{\beta}\alpha}\varphi_{\alpha\alpha_1\alpha_3 \cdots \bar{B}} + \cdots\}$$

$$= i(-1)^{p-1} \sum_{\alpha,\beta} g^{\bar{\beta}\alpha} \{\nabla_{\alpha_1}\varphi_{\alpha\alpha_2} \cdots \bar{B} - \nabla_{\alpha_2}\varphi_{\alpha\alpha_1\alpha_3} \cdots \bar{B} + \cdots\}.$$

Again by Proposition 5.2

$$(\partial\varphi)_{\alpha_0\cdots\alpha_p\bar{\beta}_1\cdots\bar{\beta}_q} = \nabla_{\alpha_0}\varphi_{\alpha_1\cdots\bar{\beta}_1\bar{\beta}_2\cdots} - \nabla_{\alpha_1}\varphi_{\alpha_0\alpha_2\cdots} + \nabla_{\alpha_2}\varphi_{\alpha_0\alpha_1\alpha_3\cdots} - \cdots$$

and

$$(\Lambda\partial\varphi)_{\alpha_1\cdots\alpha_p\bar{B}} = (-1)^p i \sum g^{\bar{\beta}\alpha}(\partial\varphi)_{\alpha\alpha_1\cdots\alpha_p\bar{\beta}B}.$$

Thus,

$$(\partial\Lambda\varphi - \Lambda\partial\varphi)_{\alpha_1\cdots\alpha_p\bar{B}} = i(-1)^{p-1}\sum_{\alpha,\beta} g^{\bar{\beta}\alpha}\nabla_\alpha\varphi_{\alpha_1\cdots\alpha_p\bar{\beta}B}$$

$$= i(\vartheta\varphi)_{\alpha_1\cdots\alpha_p\bar{\beta}_2\cdots\bar{\beta}_q}$$

by Theorem 5.2.   Q.E.D.

THEOREM 5.3.   On a Kähler manifold, $\triangle = 2\square = 2\bar{\square}$.

Proof.   (1)   We prove $\square = \bar{\square}$.

$$\square = \bar{\partial}\vartheta + \vartheta\bar{\partial} = -i\{\bar{\partial}(\partial\Lambda - \Lambda\partial) + (\partial\Lambda - \Lambda\partial)\bar{\partial}\}$$

$$= -i\{\bar{\partial}\partial\Lambda - \bar{\partial}\Lambda\partial + \partial\Lambda\bar{\partial} - \Lambda\partial\bar{\partial}\}$$

$$\bar{\square} = \partial\bar{\vartheta} - \bar{\vartheta}\partial = i\{\partial\bar{\partial}\Lambda - \bar{\partial}\Lambda\bar{\partial} + \bar{\partial}\Lambda\partial - \Lambda\bar{\partial}\partial\}$$

$$= -i\{\bar{\partial}\partial\Lambda + \partial\Lambda\bar{\partial} - \bar{\partial}\Lambda\partial - \Lambda\partial\bar{\partial}\}.$$

Thus $\bar{\square} = \square$.

(2)   Proof of $\triangle = 2\square = 2\bar{\square}$.

$$\triangle = d\delta + \delta d = (\partial + \bar{\partial})(\vartheta + \bar{\vartheta}) + (\vartheta + \bar{\vartheta})(\partial + \bar{\partial})$$

$$= \partial\bar{\vartheta} + \bar{\vartheta}\partial + \bar{\partial}\vartheta + \vartheta\bar{\partial} + \partial\vartheta + \vartheta\partial + \bar{\partial}\bar{\vartheta} + \bar{\vartheta}\bar{\partial}$$

$$= \bar{\square} + \square + \partial\vartheta + \vartheta\partial + \bar{\partial}\bar{\vartheta} + \bar{\vartheta}\bar{\partial}.$$

But

$$i(\partial\vartheta + \vartheta\partial) = \partial(\partial\Lambda - \Lambda\partial) + (\partial\Lambda - \Lambda\partial)\partial = 0$$

so

$$\partial\vartheta + \vartheta\partial = 0, \quad \bar{\partial}\bar{\vartheta} + \bar{\vartheta}\bar{\partial} = 0.$$

Thus,

$$\triangle = 2\square = 2\bar{\square}. \quad \text{Q.E.D.}$$

Now we wish to derive the important relations between $H^r(M, \mathbb{C})$ and $H^q(M, \Omega^p)$.

THEOREM 5.4.   (Hodge, Kodaira, deRham) On a compact Kähler manifold M we have

(1) $H^q(M, \Omega^p) \cong H^p(M, \Omega^q)$,

(2) $H^r(M, \mathbb{C}) \cong \bigoplus_{p+q=r} H^p(M, \Omega^q)$.

Proof.   (1) $H^q(M, \Omega^p) \cong \mathscr{H}^{p,q} = \{\varphi \mid \varphi \in \Gamma(A^{p,q}), \square\varphi = 0\}$

$\qquad\qquad H^p(M, \Omega^q) \cong \mathscr{H}^{q,p} = \{\varphi \mid \varphi \in \Gamma(A^{q,p}), \square\varphi = 0\}$.

The map $\varphi \to \bar{\varphi}$ is an antilinear isomorphism from $\Gamma(A^{p,q})$ to $\Gamma(A^{\bar{p},\bar{q}})$. If $\square\varphi = 0$, then $\overline{\square}\bar{\varphi} = 0$. But M is Kähler so $\overline{\square} = \square$, and conjugation thus gives an isomorphism $\mathscr{H}^{p,q} \cong \mathscr{H}^{q,p}$.

(2)   By the de Rham theorem,

$$H^r(M, \mathbb{C}) \cong \frac{\Gamma(dA^{r-1})}{d\Gamma(A^{r-1})}.$$

But

$$H^r(M, \mathbb{C}) \cong \mathscr{H}^r = \{\varphi \mid \varphi \in \Gamma(A^r, \triangle\varphi = 0\}. \tag{5}$$

This is the Hodge theorem. We have not proved it, but its proof is similar to that of Theorem 4.1, using de Rham's theorem and the decomposition $\mathscr{L}^q = \mathscr{H}^q + d\mathscr{L}^{q-1} + \delta\mathscr{L}^{q-1}$ analogous to that of Proposition 4.2. We leave it as an exercise to the reader. We have the following decomposition:

$$\Gamma(A^r) = \bigoplus_{p+q=r} \Gamma(A^{p,q}).$$

We claim

$$\mathscr{H}^r = \bigoplus_{p+q=r} \mathscr{H}^{p,q}. \tag{6}$$

For if $\triangle\varphi = 0$, then $\square\varphi = \tfrac{1}{2}\triangle\varphi = 0$. $\square$ maps (p, q)-forms into (p, q)-forms and thus $\square\varphi = 0$ implies $\square\varphi^{(p,q)} = 0$, where $\varphi \in \Gamma(A^r)$ and $\varphi = \sum_{p+q=r} \varphi^{(p,q)}$, $\varphi^{(p,q)} \in \Gamma(A^{p,q})$. Thus $\triangle\varphi = 0$ implies $\varphi^{(p,q)} \in \mathscr{H}^{p,q}$. So

$$\mathscr{H}^r \subseteq \bigoplus_{p+q} \mathscr{H}^{p,q}.$$

The reverse inclusion is also easy and (6) is proved. This proves Theorem 5.4. (2).   Q.E.D.

Let $h^{p,q} = h^{p,q}(M) = \dim H^q(M, \Omega^p)$ and $b_r = b_r(M) = \dim H^r(M, \mathbb{C})$; b_r is the rth Betti number of M.

COROLLARY 1.   On a compact Kähler manifold

$$b_{2p} \geq h^{p,p} \geq 1.$$

Proof. $b_{2p} \geq h^{p,p}$ is clear. Let $\omega = i \sum g_{\alpha\bar{\beta}} dz^\alpha \wedge dz^{\bar{\beta}}$ $\omega^p = \omega \wedge \cdots \wedge \omega$ (p-times). Then

$$\square \omega^p = 0,$$

because

$$\omega^p = (i)^p \sum g_{\alpha_1\bar{\beta}_1} g_{\alpha_2\bar{\beta}_2} \cdots dz^{\alpha_1} \wedge dz^{\bar{\beta}_1} \cdots$$

and

$$\nabla_\lambda g_{\alpha\bar{\beta}} = 0,\ \bar{\nabla}_\lambda g_{\alpha\bar{\beta}} = 0.$$

So, $\bar{\partial}\omega^p = 0$, $\vartheta\omega^p = 0$, by Proposition 5.3 and Theorem 5.2. Thus, $\square\omega^p = 0$. From the calculations in Theorem 1.4, we know $\omega^p \neq 0$. Thus dim $\mathcal{H}^{p,p} > 0$.

Q.E.D.

REMARK.   From Theorem 1.4 we already know $b_{2p} \geq 1$.

COROLLARY 2.   On a compact Kähler manifold

$$b_{2k+1} \equiv 0 \ (\mathrm{mod}\ 2)$$

Proof. $h^{p,q} = h^{q,p}$ so $b_{2k+1} = 2 \sum_{q=0}^{k} h^{2k+1-q,q}$.   Q.E.D.

PROPOSITION 5.5.   On a compact Kähler manifold, every holomorphic p-form φ satisfies $d\varphi = 0$.

Proof. $d\varphi = \partial\varphi + \bar{\partial}\varphi$ and $\bar{\partial}\varphi = 0$ so

$$d\varphi = \partial\varphi.$$

Thus,

$$\|\partial\varphi\|^2 = (\partial\varphi, \partial\varphi) = (\varphi, \vartheta\partial\varphi).$$

From Proposition 5.3, $\partial\Lambda - \Lambda\bar{\partial} = -i\,\bar{\vartheta}$ so

$$\|d\varphi\|^2 = -i\,(\varphi,\, \bar{\partial}\Lambda\partial\varphi - \Lambda\bar{\partial}\partial\varphi).$$

$\partial\varphi$ is holomorphic so $\bar{\partial}\partial\varphi = 0$ and Λ maps $\Gamma(A^{p,o})$ into 0. Thus $\|d\varphi\|^2 = 0$ and $d\varphi = 0$.   Q.E.D.

EXAMPLE.   We show that Kähler is needed in this proposition. Let $\mathbb{C}^3$ be the subgroup of $GL(3, \mathbb{C})$ defined by

$$\mathbb{C}^3 = \{(z_1, z_2, z_3)\} = \left\{ Z \mid Z = \begin{pmatrix} 1 & z_1 & z_2 \\ 0 & 1 & z_3 \\ 0 & 0 & 1 \end{pmatrix} \right\} \subseteq GL(3, \mathbb{C}).$$

Let

$$G = \left\{ g = \begin{pmatrix} 1 & g_1 & g_2 \\ 0 & 1 & g_3 \\ 0 & 0 & 1 \end{pmatrix} \,\Big|\, g_\lambda = m_\lambda + i\,n_\lambda;\ m_\lambda, n_\lambda \in \mathbb{Z} \right\}.$$

One easily checks that G is a subgroup of $\mathbb{C}^3$. The quotient space $\mathbb{C}^3/G$ is a *compact* complex manifold. For

$$Zg = \begin{pmatrix} 1 & z_1 + g_1 & z_2 + g_2 + g_3 z_1 \\ 0 & 1 & z_3 + g_3 \\ 0 & 0 & 1 \end{pmatrix} = Z',$$

and each point of M is represented by some $Z' = \begin{pmatrix} 1 & z_1' & z_2' \\ 0 & 1 & z_3' \\ 0 & 1 & 1 \end{pmatrix}$ with $z_\lambda' = x_\lambda' +$

$i\,y_\lambda'$ where $0 \le x_\lambda' \le 1$, $0 \le y_\lambda' \le 1$. The terms of Z' satisfy

$$z_1' = z_1 + g_1,\ z_2' = z_2 + g_2 + g_3 z_1,\ z_3' = z_3 + g_3.$$

Thus

$$dz_1' = dz_1,\ dz_3' = dz_3,\ dz_2' = dz_2 + g_3\, dz_1$$

$$= dz_2 + (z_3' - z_3)\, dz_1.$$

Hence, $\varphi = dz_2' - z_3'\, dz_1' = dz_2 - z_3\, dz_1$ is a holomorphic 1-form on $\mathbb{C}^3$ invariant under G; if $\pi : \mathbb{C}^3 \to \mathbb{C}^3/G = M$ is the canonical (holomorphic) map, there is a well-defined holomorphic 1-form ψ on M such that $\pi^*(\psi) = \varphi$. Similarly there are nonzero forms ξ, η on M such that $\pi^*(\xi) = dz_1$, $\pi^*(\eta) = dz_2$ Since $dz_3 \wedge dz_1 \ne 0$,

$$\xi \wedge \eta \ne 0.$$

But $d\varphi = dz_3 \wedge dz_1$ and hence

$$d\psi = \xi \wedge \eta \ne 0.$$

Thus the theorem is not true in general

6.   Curvatures on Kähler Manifolds

We assume throughout this section that M is compact and Kähler. We wish to find some expressions for the Laplacian in terms of curvature tensors which will be used in Section 7 in the proof of the vanishing theorems. Let $\omega = i \sum g_{\alpha\bar\beta}\, dz^\alpha \wedge dz^\beta$ be the Kähler form on the Kähler manifold M. We define

$$[\nabla_\lambda, \bar\nabla_\nu] = \nabla_\lambda \bar\nabla_\nu - \bar\nabla_\nu \nabla_\lambda,$$

the bracket of ∇_λ and $\bar\nabla_\nu$.

PROPOSITION 6.1. $[\nabla_\lambda, \bar{\nabla}_\nu]\xi^\alpha = -\sum_{\beta=1}^n R^\alpha_{\beta\bar{\nu}\lambda}\,\xi^\beta$, where

$$R^\alpha_{\beta\bar{\nu}\lambda} = \bar{\partial}_\nu \Gamma^\alpha_{\lambda\beta}.$$

DEFINITION 6.1.   The tensor field $R^\alpha_{\beta\bar{\nu}\lambda}$ is called the *curvature* of the metric ω.

Proof. (of the proposition).

$$\bar{\nabla}_\nu \xi^\alpha = \bar{\partial}_\nu \xi^\alpha,$$

so

$$\nabla_\lambda \bar{\nabla}_\nu \xi^\alpha = \partial_\lambda \bar{\partial}_\nu \xi^\alpha + \sum_\beta \Gamma^\alpha_{\lambda\beta} \bar{\partial}_\nu \xi^\beta.$$

Similarly,

$$\bar{\nabla}_\nu(\nabla_\lambda \xi^\alpha) = \bar{\partial}_\nu\left(\partial_\lambda \xi^\alpha + \sum_\beta \Gamma^\alpha_{\lambda\beta}\right)\xi^\beta$$

$$= \bar{\partial}_\nu \partial_\lambda \xi^\alpha + \sum_\beta \Gamma^\alpha_{\lambda\beta} \bar{\partial}_\nu \xi^\beta + \sum_\beta \bar{\partial}_\nu \Gamma^\alpha_{\lambda\beta} \xi^\beta.$$

Thus

$$\nabla_\lambda \bar{\nabla}_\nu - \bar{\nabla}_\nu \nabla_\lambda = -\sum_\beta \bar{\partial}_\nu \Gamma^\alpha_{\lambda\beta} \zeta^\beta. \qquad \text{Q.E.D.}$$

We wish to investigate the symmetries of $R^\alpha_{\beta\bar{\nu}\lambda}$. Let $R_{\bar{\alpha}\beta\bar{\nu}\lambda} = \sum_{\mu=1}^n g_{\mu\bar{\alpha}} R^\mu_{\beta\bar{\nu}\lambda}$. By Theorem 1.2 we can find real C^∞ K_U on each small coordinate patch $U \subset M$ such that $g_{\alpha\bar{\beta}} = \partial_\alpha\bar{\partial}_\beta K_U$ on U. We claim:

PROPOSITION 6.2.

$$R_{\bar{\alpha}\beta\bar{\nu}\lambda} = \bar{\partial}_\alpha \partial_\beta \bar{\partial}_\nu \partial_\lambda K_U - \sum_{\mu,\tau=1}^n g^{\bar{\mu}\tau}(\bar{\partial}_\mu \partial_\beta \partial_\lambda K_U)(\partial_\tau \bar{\partial}_\alpha \bar{\partial}_\nu K_U)$$

on any small open set where K_U is a C^∞ real function.

Proof. $\Gamma^\alpha_{\lambda\beta} = \sum_\tau g^{\bar{\tau}\alpha} \partial_\lambda g_{\beta\bar{\tau}}$, so

$$R^\alpha_{\beta\bar{\nu}\lambda} = \bar{\partial}_\nu \Gamma^\alpha_{\lambda\beta} = \sum_\mu g^{\bar{\mu}\alpha}\bar{\partial}_\nu \partial_\lambda g_{\beta\bar{\mu}} + \sum_\mu \bar{\partial}_\nu g^{\bar{\mu}\alpha}\partial_\lambda g_{\beta\bar{\mu}}.$$

Then

$$\sum_\alpha g_{\alpha\bar{\tau}} R^\alpha_{\beta\bar{\nu}\lambda} = \sum_{\mu,\alpha} g_{\alpha\bar{\tau}} \bar{\partial}_\nu g^{\bar{\mu}\alpha}\partial_\lambda g_{\beta\bar{\mu}} + \bar{\partial}_\nu \partial_\lambda g_{\beta\bar{\tau}}.$$

Since

$$\sum_\alpha g_{\alpha\bar{\tau}} g^{\bar{\mu}\alpha} = \delta^{\bar{\mu}}_{\bar{\tau}}, \sum_\alpha g_{\alpha\bar{\tau}} \bar{\partial}_\nu g^{\bar{\mu}\alpha} = -\sum_\alpha g^{\bar{\mu}\alpha} \bar{\partial}_\nu g_{\alpha\bar{\tau}}.$$

Thus

$$R_{\bar{\tau}\beta\bar{\nu}\lambda} = \bar{\partial}_\nu \partial_\lambda \bar{\partial}_\tau \partial_\beta K_U - \sum_{\alpha,\mu} g^{\bar{\mu}\alpha} \bar{\partial}_\nu g_{\alpha\bar{\tau}} \partial_\lambda g_{\beta\bar{\mu}}$$

$$= \bar{\partial}_\nu \partial_\lambda \bar{\partial}_\tau \partial_\beta K_U - \sum_{\alpha,\mu} g^{\bar{\mu}\alpha}(\partial_\alpha \bar{\partial}_\tau \bar{\partial}_\nu K_U)(\bar{\partial}_\mu \partial_\beta \partial_\lambda K_U). \qquad \text{Q.E.D.}$$

PROPOSITION 6.3.   (1) $R_{\bar{\alpha}\beta\bar{\nu}\lambda} = R_{\bar{\nu}\beta\bar{\alpha}\lambda} = R_{\bar{\nu}\lambda\bar{\alpha}\beta} = R_{\bar{\alpha}\lambda\bar{\nu}\beta}.$

(2) $\overline{R_{\bar{\alpha}\beta\bar{\nu}\lambda}} = R_{\bar{\beta}\alpha\bar{\lambda}\mu}.$

Proof. Proposition 6.3 (1) is clear from Proposition 6.2. For part (2), conjugate 6.2 and remember that K_U is real.   Q.E.D.

DEFINITION 6.2.   The tensor field

$$R_{\bar{\nu}\lambda} = \sum_{\alpha,\beta} g^{\bar{\alpha}\beta} R_{\bar{\alpha}\beta\bar{\nu}\lambda} = \sum_{\beta=1}^{n} R^{\beta}_{\beta\lambda\nu}$$

is called the *Ricci curvature*.

PROPOSITION 6.4. $R_{\bar{\nu}\lambda} = \partial_\lambda \bar{\partial}_\nu \log g$, where $g = \det(g_{\alpha\bar{\beta}})$.

Proof. $R_{\bar{\nu}\lambda} = \bar{\partial}_\nu \sum_\beta \Gamma^{\beta}_{\lambda\beta}$
$$= \bar{\partial}_\nu \partial_\lambda \log g \text{ by (2), Section 5 of this chapter.} \qquad \text{Q.E.D.}$$

We have a few more simple computational results.

PROPOSITION 6.5. $[\nabla_\lambda, \bar{\nabla}_\nu]\varphi_\alpha = \sum_{\beta=1}^{n} R^{\beta}_{\alpha\bar{\nu}\lambda} \varphi_\beta.$

Proof. $\bar{\nabla}_\nu \varphi_\alpha = \bar{\partial}_\nu \varphi_\alpha$, so
$$\nabla_\lambda \bar{\nabla}_\nu \varphi_\alpha = \partial_\lambda \bar{\partial}_\nu \varphi_\alpha - \sum_\beta \Gamma^{\beta}_{\lambda\alpha} \bar{\partial}_\nu \varphi_\beta.$$

Since we also have

$$\nabla_\lambda \varphi_\alpha = \partial_\lambda \varphi_\alpha - \sum_{\beta=1}^{n} \Gamma^{\beta}_{\lambda\alpha} \varphi_\beta,$$

we get,

$$\bar{\nabla}_\nu \nabla_\lambda \varphi_\alpha = \bar{\partial}_\nu(\nabla_\lambda \varphi_\alpha)$$
$$= \bar{\partial}_\nu \partial_\lambda \varphi_\alpha - \sum_\beta \bar{\partial}_\nu \Gamma^{\beta}_{\lambda\alpha} \varphi_\beta - \sum_\beta \Gamma^{\beta}_{\lambda\alpha} \bar{\partial}_\nu \varphi_\beta.$$

Thus

$$\nabla_\lambda \bar{\nabla}_\nu \varphi_\alpha - \bar{\nabla}_\nu \nabla_\lambda \varphi_\alpha = \sum_\beta R^{\beta}_{\alpha\bar{\nu}\lambda} \varphi_\beta. \qquad \text{Q.E.D.}$$

PROPOSITION 6.6. $[\nabla_\lambda, \bar{\nabla}_\nu]\varphi_{\bar{\alpha}} = -\sum_\beta R_{\bar{\alpha}}{}^{\bar{\beta}}{}_{\bar{\nu}\lambda}\varphi_{\bar{\beta}}$.

Proof. Conjugate Proposition 6.5 to get

$$[\bar{\nabla}_\lambda, \nabla_\nu]\bar{\varphi}_\alpha = \sum \overline{R_{\alpha\bar{\nu}\lambda}^\beta}\,\bar{\varphi}_\beta$$

$$= \overline{\sum_{\gamma,\beta} g^{\bar{\gamma}\beta} R_{\bar{\gamma}\alpha\bar{\nu}\lambda}\,\varphi_\beta}$$

$$= \sum_{\gamma,\beta} g^{\bar{\beta}\gamma} R_{\bar{\alpha}\gamma\bar{\lambda}\nu}\,\bar{\varphi}_\beta$$

$$= \sum_\beta R_{\bar{\alpha}}{}^{\bar{\beta}}{}_{\bar{\lambda}\nu}\,\bar{\varphi}_\beta.$$

Thus, $[\nabla_\nu, \bar{\nabla}_\lambda]\varphi_{\bar{\alpha}} = -\sum_\beta R_{\bar{\alpha}}{}^{\bar{\beta}}{}_{\bar{\lambda}\nu}\,\bar{\varphi}_\beta$.   Q.E.D.

We could similarly prove

$$[\nabla_\lambda, \bar{\nabla}_\nu]\zeta^\alpha{}_{\beta\bar{\gamma}} = -\sum_\tau R^\alpha{}_{\tau\bar{\lambda}\nu}\zeta^\tau{}_{\beta\bar{\gamma}} + \sum_\tau R^\tau{}_{\beta\bar{\lambda}\nu}\zeta^\alpha{}_{\tau\bar{\gamma}}$$

$$+ \sum_\tau R^{\bar{\tau}}{}_{\bar{\gamma}\bar{\lambda}\nu}\zeta^\alpha{}_{\beta\bar{\tau}}.$$

THEOREM 6.1. For any (p,q)-form $\varphi = 1/p!\,q!\sum \varphi_{\alpha_1\cdots\bar{\beta}_q}\,dz^{\alpha_1}\wedge\cdots\wedge dz^{\bar{\beta}_q}$.

$$(\Box\varphi)_{\alpha_1\cdots\bar{\beta}_q} = -\sum_{\alpha,\beta} g^{\bar{\beta}\alpha}\nabla_\alpha\bar{\nabla}_\beta\,\varphi_{\alpha_1\cdots\bar{\beta}_q}$$

$$+ \sum_{i=1}^{p}\sum_{k=1}^{q}\sum_{\tau,\sigma} R^\tau{}_{\alpha_i\bar{\beta}_k}{}^{\bar{\sigma}}\,\varphi_{\alpha_1\cdots\alpha_{i-1}\tau\alpha_{i+1}\cdots\bar{\beta}_{k-1}\bar{\sigma}\bar{\beta}_{k+1}\cdots\bar{\beta}_q}$$

$$- \sum_{k=1}^{q}\sum_\tau R_{\bar{\beta}_k}{}^{\bar{\tau}}\,\varphi_{\alpha_1\cdots\bar{\beta}_{k-1}\bar{\tau}\bar{\beta}_{k+1}\cdots\bar{\beta}_q},$$

where $R_\beta{}^\tau = \sum_\lambda g^{\bar{\tau}\lambda} R_{\bar{\beta}\lambda}$.

Proof. As usual let A denote $\alpha_1\cdots a_p$. Then

$$(\bar{\partial}\varphi)_{A\bar{\beta}_0\cdots\bar{\beta}_q} = (-1)^p\{\bar{\nabla}_{\beta_0}\varphi_{A\bar{\beta}_1\bar{\beta}_2\cdots} - \bar{\nabla}_{\beta_1}\varphi_{A\bar{\beta}_0\bar{\beta}_2\cdots} + \cdots\}$$

$$= (-1)^p\sum_{\lambda=0}^{q}(-1)^\lambda\bar{\nabla}_{\beta\lambda}\varphi_{A\bar{\beta}_0\cdots\hat{\bar{\beta}}_\lambda\cdots\bar{\beta}_q}$$

and

$$(\partial\bar{\partial}\varphi)_{A\bar{\beta}_1\cdots\bar{\beta}_q} = -(-1)^p\sum_{\alpha,\beta} g^{\bar{\beta}\alpha}\nabla_\alpha(\bar{\partial}\varphi)_{A\bar{\beta}\bar{\beta}_1\cdots\bar{\beta}_q}$$

$$= -\sum_{\alpha,\beta}\left(g^{\bar{\beta}\alpha}\nabla_\alpha\bar{\nabla}_\beta\varphi_{A\bar{\beta}_1\cdots\bar{\beta}_q} + \sum_{\lambda=1}^{q}(-1)^{\lambda+1}g^{\bar{\beta}\alpha}\nabla_\alpha\bar{\nabla}_{\bar{\beta}_\lambda}\varphi_{A\bar{\beta}\cdots\hat{\bar{\beta}}_\lambda\cdots\bar{\beta}_q}\right).$$

Also

$$(\vartheta\varphi)_{A\bar{\beta}_2\cdots\bar{\beta}_2} = -(-1)^p \sum_{\alpha,\beta} g^{\bar{\beta}\alpha}\nabla_\alpha \varphi_{A\bar{\beta}\bar{\beta}_2\cdots\bar{\beta}_q}$$

so

$$(\bar{\partial}\vartheta\varphi)_{A\bar{\beta}_1\cdots\bar{\beta}_q} = -\sum_{\lambda=1}^{q}(-1)^{\lambda+1}\bar{\nabla}_{\beta_\lambda}(g^{\bar{\beta}\alpha}\nabla_\alpha \varphi_{A\bar{\beta}\bar{\beta}_1\cdots\hat{\bar{\beta}}_\lambda\cdots\bar{\beta}_q}).$$

Thus

$$(\Box\varphi)_{A\bar{\beta}_1\cdots\bar{\beta}_q} = (\vartheta\bar{\partial}+\bar{\partial}\vartheta)\varphi_{A\bar{\beta}_1\cdots\bar{\beta}_q}$$

$$= -\sum_{\alpha,\beta=1}^{n} g^{\bar{\beta}\alpha}\nabla_\alpha\bar{\nabla}_\beta\,\varphi_{A\bar{\beta}_1\cdots\bar{\beta}_q} \tag{1}$$

$$-\sum_{\lambda=1}^{q}(-1)^\lambda g^{\bar{\beta}\alpha}[\nabla_\alpha,\bar{\nabla}_{\beta_\lambda}]\varphi_{A\bar{\beta}\cdots\hat{\bar{\beta}}_\lambda\cdots\bar{\beta}_q}.$$

Let us calculate the second term on the right-hand side of (1). For a form φ of type (1, 0)

$$[\nabla_\lambda,\bar{\nabla}_\nu]\varphi_\alpha = \sum_\tau R^\tau{}_{\alpha\bar{\nu}\lambda}\,\varphi_\tau$$

and

$$\sum_\lambda g^{\bar{\beta}\lambda}[\nabla_\lambda,\bar{\nabla}_\nu]\varphi_\alpha = \sum_\tau R^\tau{}_{\alpha\bar{\nu}}{}^{\bar{\beta}}\,\varphi_\tau.$$

For a form φ of type (0, 1)

$$[\nabla_\lambda,\bar{\nabla}_\nu]\varphi_{\bar{\beta}} = -\sum_\tau R_{\bar{\beta}}{}^{\bar{\tau}}{}_{\bar{\nu}\lambda}\,\varphi_{\bar{\tau}}$$

, so

$$\sum_\lambda g^{\bar{\nu}\lambda}[\nabla_\lambda,\bar{\nabla}_\nu]\varphi_{\bar{\beta}} = -\sum R_{\bar{\beta}}{}^{\bar{\tau}}{}_{\bar{\nu}}{}^{\bar{\nu}}\varphi_{\bar{\tau}}.$$

We also have

$$\sum_\beta R_{\bar{\beta}}{}^{\bar{\tau}}{}_{\bar{\beta}_1}{}^{\bar{\beta}} = \sum_{\alpha,\beta,\gamma} g^{\bar{\tau}\alpha}g^{\bar{\beta}\gamma}R_{\beta\alpha\bar{\beta}_1\gamma} \tag{2}$$

$$= \sum_\alpha g^{\bar{\tau}\alpha}R_{\bar{\beta}_1\alpha} = R_{\bar{\beta}_1}{}^{\bar{\tau}}.$$

Similarly, we see

$$\sum_{\alpha,\beta} g^{\bar{\beta}\alpha}[\nabla_\alpha,\bar{\nabla}_{\beta_\lambda}]\varphi_{A\bar{\beta}\bar{\beta}_1\cdots\hat{\bar{\beta}}_\lambda\cdots\bar{\beta}_q}$$

$$= \sum_{\tau,\beta}\left(\sum_{i=1}^{p} R^\tau{}_{\alpha_i\bar{\beta}_\lambda}{}^{\bar{\beta}}\varphi_{\alpha_1\cdots(\tau)_i\cdots\alpha_p\bar{\beta}\bar{\beta}_1\cdots\hat{\bar{\beta}}_\lambda\cdots\bar{\beta}_q}\right.$$
$$\uparrow i^{th}\ \text{place}$$

$$\left.-R_{\bar{\beta}}{}^{\bar{\tau}}{}_{\bar{\beta}_\lambda}{}^{\bar{\beta}}\varphi_{\alpha_1\cdots\alpha_p\bar{\tau}\bar{\beta}_1\cdots\hat{\bar{\beta}}_\lambda\cdots\bar{\beta}_q}\right)$$

$$-\sum_{\bar{\tau},\beta}\left(\sum_{k=1}^{q} R_{\bar{\beta}_k}{}^{\bar{\tau}}{}_{\bar{\beta}_\lambda}{}^{\bar{\beta}}\varphi_{\alpha_1\cdots\alpha_p\bar{\beta}\bar{\beta}_1\cdots(\bar{\tau})_i\cdots\hat{\bar{\beta}}_\lambda\cdots\bar{\beta}_q}\right).$$
$$\uparrow k^{th}\ \text{place}$$

Since $R_{\bar{\beta}_k{}^\tau{}_{\bar{\beta}_\lambda}{}^{\bar{\beta}}}$ is symmetric in $\tau\bar{\beta}$ and φ is antisymmetric in $\tau\bar{\beta}$, the last term is zero. Thus, using (2)

$$\sum_{\alpha,\beta} g^{\bar{\beta}\alpha}[\nabla_\alpha, \bar{\nabla}_{\bar{\beta}_\lambda}]\varphi_{A\beta\bar{\beta}_1\cdots\hat{\bar{\beta}}_\lambda\cdots\bar{\beta}_\lambda} \tag{3}$$

$$= \sum_{i=1}^{p}\sum_{\tau,\beta} R^\tau{}_{\alpha_i\bar{\beta}_\lambda}{}^{\bar{\beta}}\varphi_{\alpha_1\cdots(\tau)_i\cdots\alpha_p\bar{\beta}\cdots\hat{\bar{\beta}}_\lambda\cdots\bar{\beta}_q}$$

$$- \sum_\tau R_{\bar{\beta}_\lambda}{}^{\bar{\tau}}\varphi_{\alpha_1\cdots\alpha_p\tau\bar{\beta}_1\cdots\hat{\bar{\beta}}_\lambda\cdots\bar{\beta}_q},$$

where $(\tau)_i$ means that τ occurs in the ith place. Multiplying (2) by $(-1)^\lambda$ and plugging into (1) yields the theorem.   Q.E.D.

We want to derive a similar theorem for $\square_a$ acting on $\Gamma(A^{p,q}(F))$ where F is a complex line bundle defined by the 1-cocycle $\{f_{jk}\}$. (As usual in this section we are assuming that M is Kähler.) A form $\varphi \in \Gamma(A^{p,q}(F))$ is given locally by a family of (p,q)-forms $\{\varphi_j\}$ on $\{U_j\}$ where $\{U_j\}$ is a covering of M with coordinate patches over which F is trivial such that

$$\varphi_j = f_{jk}\varphi_k \text{ on } U_j \cap U_k. \tag{4}$$

Let $\omega = i\sum g_{\alpha\bar{\beta}}\,dz^\alpha \wedge dz^\beta$ be the Kähler form of M and suppose we have chosen an Hermitian form $\langle,\rangle$ on the fibres, so

$$\langle\zeta,\zeta\rangle = a_j|\zeta_j|^2,$$

where ζ_j is a fibre coordinate of ζ and $a_j(z)$ is a real positive C^∞ function on U_j. Then $a_j|\zeta_j|^2 = a_k|\zeta_k|^2$ implies

$$|f_{jk}|^2 = \frac{a_k}{a_j}. \tag{5}$$

For two forms $\varphi, \psi \in \Gamma(A^{p,q}(F))$ their inner product (φ, ψ) is then

$$(\varphi, \psi) = \int_M a_j\varphi_j \wedge * \bar{\psi}_j.$$

The integrand is well defined since by (4) and (5),

$$a_j\varphi_j \wedge * \bar{\psi}_j = a_k\varphi_k \wedge * \bar{\psi}_k \text{ on } U_j \cap U_k.$$

Recall that we defined

$$(\bar{\partial}\varphi)_j = \bar{\partial}\varphi_j$$

and by Equation (1), Section 3

$$(\vartheta_a\varphi)_j = \vartheta\varphi_j - *(a_j^{-1}\partial a_j \wedge * \varphi_j)$$

$$= \vartheta\varphi_j - *\left(\sum a_j^{-1}\partial_\alpha a_j\,dz^\alpha \wedge * \varphi_j\right)$$

$$= -*(\partial * \varphi_j) - *\left(\sum a_j^{-1}\partial_\alpha a_j\,dz^\alpha \wedge * \varphi_j\right).$$

Recall also by Proposition 5.2 and Theorem 5.2

$$\vartheta \varphi_j = - * (\partial * \varphi_j) = - * \left(\sum_{\alpha=1}^{n} \nabla_\alpha \, dz^\alpha \wedge * \varphi_j \right)$$

and

$$(\vartheta \varphi_j)_{\alpha_1 \cdots \alpha_p \bar\beta_1 \cdots \bar\beta_{q-1}} = - \sum_\beta g^{\bar\beta\alpha} \nabla_\alpha \varphi_{j\bar\beta\alpha_1 \cdots \alpha_p \bar\beta_1 \cdots \bar\beta_{q-1}}.$$

Thus we have:

PROPOSITION 6.7.

$$(\vartheta_\alpha \varphi)_{\alpha_1 \cdots \bar\beta_1 \cdots \bar\beta_{q-1}} = -(-1)^p \sum_{\alpha, \beta} g^{\bar\beta\alpha} (\nabla_\alpha + \partial_\alpha \log a_j) \varphi_{j\alpha_1 \cdots \alpha_p \bar\beta\bar\beta_1 \cdots \bar\beta_{q-1}}.$$

Proof. We need only show

$$- * \left(\sum_\alpha a_j^{-1} \partial_\alpha a_j \, dz^\alpha \wedge * \varphi_j \right)_{\alpha_1 \cdots \alpha_p \bar\beta_2 \cdots \bar\beta_{q-1}} = - \sum_{\alpha, \beta} g^{\bar\beta\alpha} a_j^{-1} \partial_\alpha a_j \, \varphi_{j\bar\beta\alpha_1 \cdots \alpha_p \bar\beta_1 \cdots}.$$

To do this we assume $g^{\bar\beta\alpha}(z_0) = \delta^{\bar\beta\alpha}$ and $\varphi_j = dz^{A_p} \wedge dz^{\bar\beta_q}$. Then

$$* \varphi_j = \mathrm{sgn} \begin{pmatrix} B_q \, B_{n-q} \\ A_p \, A_{n-p} \end{pmatrix} i^n (-1)^{\frac{1}{2}n(n-1)+np} \, dz^{B_{n-q}} \wedge dz^{\bar A_{n-p}}.$$

Next

$$* (dz^\alpha \wedge dz^{B_{n-q}} \wedge dz^{\bar A_{n-p}}) = \eta \, dz^{A_p} \wedge dz^{\bar X},$$

where η and X are as follows: X is the increasing set of numbers $(x_1 \cdots x_{q-1})$ complementary to the set $\alpha B_{n-q} \subseteq (1 \cdots n)$. If we order αB_{n-q} in increasing order Y let

$$\varepsilon = \mathrm{sgn} \begin{pmatrix} Y \\ \alpha B_{n-q} \end{pmatrix}.$$

Then

$$\eta = \varepsilon i^n (-1)^{\frac{1}{2}n(n-1)+n(n-q+1)} \, \mathrm{sgn} \begin{pmatrix} A_{n-p} \, A_p \\ Y \quad X \end{pmatrix}$$

$$= i^n (-1)^{\frac{1}{2}n(n-1)+n(n-q+1)} \, \mathrm{sgn} \begin{pmatrix} A_{n-p} \, A_p \\ \alpha B_{n-q} \, X \end{pmatrix}.$$

Thus,

$$* (dz^\alpha \wedge * \varphi_j) = (-1)^p \begin{pmatrix} B_q \\ \alpha X \end{pmatrix} dz^{A_p} \wedge dz^{\bar X}.$$

Hence,

$$* \left(\sum_{\alpha=1}^{n} c_\alpha \, dz^\alpha \wedge * \varphi_j \right) = (-1)^p \sum_{\alpha \in B_q} c_\alpha \, \mathrm{sgn} \begin{pmatrix} B_q \\ \alpha X \end{pmatrix} dz^{A_p} \wedge dz^{\bar X},$$

where $X = (x_1 \cdots x_{q-1})$ is an increasing set of numbers from $(1 \cdots n)$ so that X is the complement of α in B_q. But we also have

$$(-1)^p \sum_{\alpha \cup X = B_q} c_\alpha \, \varphi_{jA_p \bar{\alpha}\bar{x}_1 \cdots \bar{x}_{q-1}} \, dz^{A_p} \wedge dz^X$$

$$= (-1)^p \sum_{\alpha \in B_q} c_\alpha \, \mathrm{sgn} \begin{pmatrix} B_q \\ \alpha X \end{pmatrix} dz^{A_p} \wedge dz^X$$

corresponding to the right-hand side of (6). Letting $c_\alpha = a_j^{-1} \partial_\alpha a_j$ we see (6) is verified at z_0 in the case $g^{\bar{\beta}\alpha}(z_0) = \delta^{\bar{\beta}\alpha}$, $\varphi_j = dz^{A_p} \wedge dz^{\beta_q}$. The general case follows easily from this.   Q.E.D.

We want to define covariant differentiation of sections $\varphi \in \Gamma(A^{p,q}(F))$. Let $\varphi = \{\varphi_j\} \in \Gamma(A^{(p,q)}(F))$. Then $\varphi_{jA\bar{B}} = f_{jk}\varphi_{kA\bar{B}}$. One can easily check the following fact:

If φ is a form and f is a C^∞ function,

$$\bar{\nabla}_\alpha(f \cdot \varphi) = (\bar{\partial}_\alpha f)\varphi + f \cdot \bar{\nabla}_\alpha \varphi$$

and

$$\nabla_\alpha(f \cdot \varphi) = (\partial_\alpha f)\varphi + f \cdot \nabla_\alpha \varphi.$$

But in our case $\bar{\partial} f_{jk} = 0$ so

$$\bar{\nabla}_\alpha \varphi_{jAB} = f_{jk} \bar{\nabla}_\alpha \varphi_{kAB}.$$

Thus we define

$$\bar{\nabla}_\alpha \varphi = \{\bar{\nabla}_\alpha \varphi_j\}. \tag{7}$$

However,

$$\nabla_\alpha \varphi_{jA\bar{B}} = f_{jk} \nabla_\alpha \varphi_{kAB} + \partial_\alpha f_{jk} \, \varphi_{kAB}$$

so we must make a different definition of ∇_α which depends on a_j. We know

$$\frac{a_k}{a_j} = |f_{jk}|^2 = f_{jk}\bar{f}_{jk}.$$

Hence,

$$a_j \varphi_{jA\bar{B}} = \frac{1}{\bar{f}_{jk}} a_k \varphi_{kA\bar{B}}.$$

Now

$$\nabla_\alpha\left(\frac{1}{\bar{f}_{jk}}\right) = \partial_\alpha\left(\frac{1}{\bar{f}_{jk}}\right) = 0,$$

so that

$$\nabla_\alpha(a_j \varphi_{jAB}) = \left(\frac{1}{\bar{f}_{jk}}\right)(\nabla_\alpha a_k \varphi_{kAB}),$$

thus proving

$$\frac{1}{a_j}\nabla_\alpha(a_j\,\varphi_{jA\bar B}) = f_{jk}\left(\frac{1}{a_k}\nabla_\alpha(a_k\,\varphi_{kA\bar B})\right).$$

We define

$$\nabla_\alpha^{(a)}\varphi = \left\{\frac{1}{a_j}\nabla_\alpha(a_j\,\varphi_j)\right\} \tag{8}$$

$$= \{(\nabla_\alpha + \partial_\alpha\log a_j)(\varphi_j)\}.$$

Recall $\square_a = \bar\partial\vartheta_a + \vartheta_a\bar\partial$.

THEOREM 6.2.

$$\square_a\,\varphi_{jA_p\bar B_q} = -\sum g^{\bar\beta\alpha}\nabla_\alpha^{(a)}\bar\nabla_\beta\,\varphi_{jA_p\bar B_q}$$

$$+ \sum_{k=1}^q\sum_\tau (X^{\bar\tau}{}_{\bar\beta_k} - R_{\bar\beta_k}{}^{\bar\tau})\varphi_{jA_p\bar B_q\cdots(\bar\tau)_k\cdots\bar\beta_q}$$

$$\times \sum_i\sum_k\sum_{\tau,\bar\sigma} R^\tau{}_{\alpha_i\bar\beta_k}{}^{\bar\sigma}\,\varphi_{j\alpha_1\cdots(\bar\tau)_i\cdots\alpha_p\bar\beta_1\cdots(\bar\sigma)_k\cdots\bar\beta_q},$$

where $X_{\bar\tau}^{\bar\beta} = -\bar\nabla_\beta\xi^{\bar\tau}$ and $\xi^{\bar\tau} = \sum_\alpha g^{\bar\tau\alpha}\partial_\alpha\log a_j$.

Proof. We omit the subscript j from our calculations. From Proposition 6.7,

$$(\vartheta_a\varphi)_{A_p\bar B_{q-1}} = (\vartheta\varphi)_{A_p\bar B_{q-1}} - (-1)^p\sum_\beta \xi^\beta\varphi_{A_p\bar\beta\bar B_{q-1}}.$$

We denote the last term in this expression by $(\xi\varphi)_{A_p\bar B_{q-1}}$. Then

$$(\square_a\varphi)_{A_p\bar B_q} = [(\bar\partial\vartheta_a + \vartheta_a\bar\partial)\varphi]_{A_p\bar B_q}$$

$$= (\square\varphi)_{A_p\bar B_q} + [(\bar\partial\xi + \xi\bar\partial)\varphi]_{A_p\bar B_q}.$$

Computing gives (using Proposition 5.3),

$$[\bar\partial(\xi\varphi)]_{A_p\bar B_q} = (-1)^p\sum_{k=1}^q (-1)^{k-1}\bar\nabla_{\beta_k}(\xi\varphi)_{A_p\bar\beta_1\cdots\hat{\bar\beta}_k\cdots\bar\beta_q}$$

$$= -\sum_{k=1}^q (-1)^{k-1}\bar\nabla_{\beta_k}\sum_\beta \xi^{\bar\beta}\varphi_{A_p\bar\beta\bar\beta_1\cdots\hat{\bar\beta}_k\cdots\bar\beta_q}$$

$$= -\sum_{k=1}^q\sum_\beta \bar\nabla_{\beta_k}(\xi^\beta\varphi_{A_p\bar\beta_1\cdots(\bar\beta)_k\cdots\bar\beta_q}).$$

We also have

$$(\xi\bar\partial\varphi)_{A_p\bar B_q} = -\sum_\beta\left(\xi^{\bar\beta}\bar\nabla_\beta - \sum_{k=1}^q \xi^{\bar\beta}\bar\nabla_{\beta_k}\right)\varphi_{A_p\bar\beta_1\cdots(\bar\beta)_k\cdots\bar\beta_q}.$$

So

$$(\Box_a \varphi)_{A_p \bar{B}_q} = (\Box \varphi)_{A_p \bar{B}_q} - \sum_\beta (\xi^{\bar{\beta}} \bar{\nabla}_\beta) \varphi_{A_p \bar{B}_q} - \sum_\beta \sum_{k=1}^q (\bar{\nabla}_{\beta_k} \xi^{\bar{\beta}}) \varphi_{A_p \cdots (\bar{\beta})_k \cdots} . \tag{9}$$

We now use Theorem 6.1. First note that

$$-\sum g^{\bar{\beta}\alpha} \nabla_\alpha \bar{\nabla}_\beta - \xi^{\bar{\beta}} \bar{\nabla}_\beta = -\sum g^{\bar{\beta}\alpha} (\nabla_\alpha + \partial_\alpha \log a_j) \bar{\nabla}_\beta \tag{10}$$

$$= -\sum g^{\bar{\beta}\alpha} \nabla_\alpha^{(a)} \bar{\nabla}_\beta .$$

Now use (9), (10), and Theorem 6.1 to finish the proof.   Q.E.D.

REMARK.
$$\begin{aligned}
X_{\lambda \bar{\mu}} = \sum g_{\lambda \bar{\tau}} X^{\bar{\tau}}_{\bar{\mu}} &= -\bar{\nabla}_{\bar{\mu}} \partial_\lambda \log a_j \\
&= -\partial_\lambda \bar{\partial}_{\bar{\mu}} \log a_j
\end{aligned} \tag{11}$$

is called the *curvature* of the metric a.

7. Vanishing Theorems

We wish to use the computations of Section 6 to show that the cohomology groups $H^q(M, \mathcal{O}(F))$ must vanish under certain circumstances. The first result to this effect is the following theorem. The technique is due to Bochner.

THEOREM 7.1.   If the Hermitian matrix $X_{\tau\bar{\sigma}} - R_{\bar{\sigma}\tau}$ is positive definite at each point of the Kähler manifold M, then

$$H^q(M, \mathcal{O}(F)) = 0 \qquad \text{for } q \geq 1.$$

Proof.   We first prove:

LEMMA 7.1.   Let $\omega = i \sum g_{\alpha\bar{\beta}} \, dz^\alpha \wedge dz^{\bar{\beta}}$ be the Kähler form on M. Let Φ be any 1-form on M. Then

$$\int_M \delta\Phi \, \frac{\omega^n}{n!} = 0.$$

Proof. $\varphi \wedge *\bar{\psi} = (\varphi, \psi)(z)(\omega^n/n!)$ from Theorem 2.1. Let $\varphi = 1$, $\bar{\psi} = f$ be a differentiable function. Then we see that $*f = f(z)(\omega^n/n!)$. Recall that $\int_M d\Psi = 0$ for any $(2n - 1)$-form Ψ. Thus,

$$0 = \int_M d(*\Phi) = \pm \int_M **d(*\Phi) = \pm \int_M *\delta\Phi = \pm \int \delta\Phi \cdot (\omega^n/n!). \qquad \text{Q.E.D.}$$

Now for the proof of the theorem. Recall that

$$H^q(M, \mathcal{O}(F)) \cong \mathcal{H}^{o,q}(F) = \{\varphi \mid \varphi \in \Gamma(A^{o,q}(F)), \square_a \varphi = 0\}.$$

Thus we want to show that any $\varphi \in \Gamma(A^{o,q}(F))$ satisfying $\square_a \varphi = 0$ is necessarily zero. Let

$$\Phi = a_j \sum_{\beta, \beta_q} \bar\nabla_\beta \varphi_{j\bar B_q} \varphi_j^{B_q} \, dz^{\bar\beta}.$$

Then Φ is a $(0, 1)$-form so

$$0 = \int_M \delta\Phi \cdot \frac{\omega^n}{n!} = \int_M \vartheta\Phi \cdot \frac{\omega^n}{n!}$$

since

$$\delta\Phi = \vartheta\Phi + \bar\vartheta\Phi \text{ and } \bar\vartheta : \Gamma(A^{o,1}) \to \Gamma(A^{-1,o}) = 0$$

is the zero map. Thus,

$$0 = \int \sum g^{\bar\beta\alpha}\nabla_\alpha\{a_j \sum \bar\nabla_\beta \varphi_{j\bar B_q} \cdot \overline{\varphi_j^{B_q}}\} \frac{\omega^n}{n!} \text{ (by Theorem 5.2)}$$

$$= \int a_j\{\sum g^{\bar\beta\alpha} \sum \nabla_\alpha^{(a)}\bar\nabla_\beta \varphi_{j\bar B_q} \cdot \overline{\varphi_j^{B_q}}\} \frac{\omega^n}{n!}$$

$$+ \int a_j\{\sum g^{\bar\beta\alpha} \sum \bar\nabla_\beta \varphi_{j\bar B_q} \overline{\nabla_\alpha \varphi_j^{B_q}}\} \frac{\omega^n}{n!}.$$

The term in the braces in the last term is always nonnegative since $g_{\alpha\bar\beta}$ and $g^{\bar\beta\alpha}$ are positive definite. Thus the second integral is nonnegative and the first integral is nonpositive. It is clear from the derivation of Equation (3) in Section 6 that for a (o, q)-form φ using Theorem 6.2

$$\square_a \varphi_{j\bar B_q} = -\sum g^{\bar\beta\alpha}\nabla_\alpha^{(a)}\bar\nabla_\beta \varphi_{j\bar B_q} + \sum_{k=1}^{q} \sum_\tau (X^{\bar\tau}_{\bar\beta_k} - R_{\bar\beta_k}^{\bar\tau})\varphi_{j\bar\beta 1 \cdots (\bar\tau)_k \cdots \bar\beta_q}.$$

Thus,

$$0 \geq \int a_j \sum_{B_q} \sum_{\sigma, \tau} q(X^{\bar\tau}_{\bar\sigma} - R_{\bar\sigma}^{\bar\tau})\varphi_{j\bar\beta_1 \cdots (\bar\tau)_k \cdots \bar\beta_q} \cdot \overline{\varphi^{\beta_1 \cdots (\sigma)_k \cdots \beta_q}} \frac{\omega^n}{n!}$$

since $\square_a \varphi_j = 0$. Recalling the definition of $\varphi_j^{\bar B_q}$ we see

$$0 \geq \int a_j q \sum_{A_{q-1}, B_{q-1}} \sum_{\sigma, \tau} (X_{\tau\bar\sigma} - R_{\bar\sigma\tau})g_{\alpha_1\bar\beta_1} \cdots g_{\alpha_{q-1}\bar\beta_{q-1}} \varphi^{\tau A_{q-1}} \cdot \overline{\varphi^{\sigma B_{q-1}}} \frac{\omega^n}{n!}.$$

By assumption $(X_{\tau\bar\sigma} - R_{\bar\sigma\tau})$ is positive definite. Hence, $\varphi^{\tau\alpha_1 \cdots \alpha_{q-1}} = 0$ and $\mathcal{H}^{(o,q)}(F) = 0$ for $q \geq 1$.   Q.E.D.

REMARK.   For $q = 0$ the theorem says nothing.

We now discuss the meaning of the curvature $X_{\lambda\bar\mu}$. Recall the sequence

$$0 \longrightarrow \mathbb{Z} \longrightarrow \mathcal{O} \longrightarrow \mathcal{O}^* \longrightarrow 0$$

in Section 5, Chapter 2. We get the exact cohomology sequence

$$\cdots \longrightarrow H^1(M, \mathcal{O}) \longrightarrow H^1(M, \mathcal{O}^*) \overset{\delta^*}{\longrightarrow} H^2(M, \mathbb{Z}) \longrightarrow \cdots$$

and we defined $c(F) = \delta*(F)$. Since $\mathbb{Z} \subset \mathbb{C}$ we map $H^2(M, \mathbb{Z}) \to H^2(M, \mathbb{C})$ and send $c(F) \to c(F)_{\mathbb{C}}$. De Rham's theorem then says

$$H^2(M, \mathbb{C}) \cong \frac{\Gamma(dA^1)}{d\Gamma(A^1)}.$$

THEOREM 7.2. The de Rham cohomology class of $c(F)_{\mathbb{C}}$ is represented by $(1/2\pi i) \sum X_{\lambda\bar\nu}\, dz^\lambda \wedge dz^{\bar\nu}$.

Proof. This is an exercise in tracing the de Rham isomorphism map. Let $F = \{f_{jk}\}$. Then $c(F) = [\{c_{ijk}\}]$, where

$$c_{ijk} = \frac{1}{2\pi i} \{\log f_{ij} + \log f_{jk} + \log f_{ki}\}.$$

We wish to find $\gamma \in \Gamma(dA^1)$ representing $c(F)_{\mathbb{C}}$. We can find C^∞ 1-forms σ_j on U_j such that

$$\frac{1}{2\pi i}\, d \log f_{jk} = \sigma_k - \sigma_j.$$

Then $\gamma = d\sigma_j = d\sigma_k$. Remember that $|f_{jk}|^2 = a_k/a_j$ and $X_{\lambda\bar\mu} = -\partial_\lambda \bar\partial_\mu \log a_j$. Thus

$$\log f_{jk} + \overline{\log f_{jk}} = \log a_k - \log a_j$$

and

$$d \log f_{jk} = \partial \log f_{jk} = \partial \log a_k - \partial \log a_j.$$

Let

$$\sigma_k = \frac{1}{2\pi i}\, \partial \log a_k.$$

Then

$$\gamma = d\sigma_k = \frac{1}{2\pi i}\, \bar\partial\, \partial \log a_k = -\frac{1}{2\pi i} \sum \partial_\lambda \bar\partial_\mu \log a_k\, dz^\lambda \wedge dz^{\bar\mu}$$

$$= \frac{1}{2\pi i} \sum X_{\lambda\bar\mu}\, dz^\lambda \wedge dz^{\bar\mu}. \qquad \text{Q.E.D.}$$

The line bundle over M defined by the 1-cocycle

$$K = \{J_{jk}\}, \quad J_{jk} = \frac{\partial(z_k^1 \cdots z_k^n)}{\partial(z_j^1 \cdots z_j^n)},$$

with respect to a coordinate covering $\{U_j\}$ of M, is called the canonical bundle K of M. Then it is easy to see that

$$\frac{g_j}{g_k} = |J_{jk}|^2.$$

The first Chern class $c_1(M)$ of M is then (this may be taken as a definition)

$$c_1(M) = -c(K).$$

THEOREM 7.3. The de Rham cohomology class of $c_1(M)_{\mathbb{C}}$ is

$$\left[\frac{1}{2\pi i} \sum R_{\bar{\mu}\lambda}\, dz^\lambda \wedge dz^{\bar{\mu}} \right].$$

Proof. Left to the reader.

THEOREM 7.4. Let M be a Kähler manifold. Let $F = \{f_{jk}\}$ be a line bundle. If a $(1, 1)$-form

$$\gamma = \frac{1}{2\pi i} \sum \gamma_{\lambda\bar{\mu}}\, dz^\lambda \wedge dz^{\bar{\mu}}$$

is real (that is, $\bar{\gamma} = \gamma$), $d\gamma = o$, and if $[\gamma] = c(F)$, then there exists $\{a_j\}$, a_j C^∞ functions on U_j, $a_j > 0$ satisfying $a_j |f_{jk}|^2 = a_k$ such that

$$\gamma = \frac{i}{2\pi} \partial \bar{\partial} \log a_j,$$

that is, $\gamma_{\lambda\bar{\mu}} = -\partial_\lambda \partial_{\bar{\mu}} \log a_j$.

Proof. Choose any metric $\hat{a} = \{\hat{a}_j\}$ on F. That is, $\hat{a}_j \in C^\infty(U_j)$, $\hat{a}_j > 0$ and $\hat{a}_j |f_{jk}|^2 = \hat{a}_k$. Then define

$$\xi = \frac{1}{2\pi i} \sum X_{\lambda\bar{\mu}}\, dz^\lambda \wedge dz^{\bar{\mu}},$$

where

$$X_{\lambda\bar{\mu}} = -\partial_\lambda \partial_{\bar{\mu}} \log \hat{a}_k;$$

that is,

$$\xi = \frac{i}{2\pi} \partial \bar{\partial} \log \hat{a}_j.$$

Then as in Theorem 7.2 $[\xi] = c(F)_{\mathbb{C}}$ so $\xi - \gamma = d\varphi$, where φ is a 1-form. Thus, $d\varphi$ is a $(1, 1)$-form and

$$d\varphi = \xi - \gamma = \eta + \Box\psi,$$

where η and ψ are $(1, 1)$-forms and $\Box\eta = 0$. But then

$$\Delta\eta = 2\,\Box\eta = 0 \qquad \text{so} \qquad d\eta = \delta\eta = 0.$$

Also $dd\varphi = 0$ so

$$d\varphi = \eta + \tfrac{1}{2}(d\delta + \delta d)\,\psi$$

implies

$$0 = d\delta d\psi \tag{1}$$

and hence

$$(\delta d\psi, \delta d\psi) = (d\psi, d\delta d\psi) = 0.$$

Thus,

$$\delta d\psi = 0$$

so

$$d\varphi = \eta + \tfrac{1}{2}\,d\delta\psi.$$

Then

$$\begin{aligned}
(\eta, d\varphi) = (\delta\eta, \varphi) &= 0 \\
&= (\eta, \eta) + \tfrac{1}{2}\,(\eta, d\delta\psi) \\
&= (\eta, \eta).
\end{aligned}$$

Hence $\eta = 0$. Using Equation (1)

$$0 = (\delta d\psi, \psi) = (d\psi, d\psi)$$

so

$$d\psi = 0. \tag{2}$$

From (2)

$$0 = d\psi = \partial\psi + \bar\partial\psi.$$

Thus, $\partial\psi = \bar\partial\psi = 0$ since ψ is of type $(1, 1)$. So

$$\begin{aligned}
d\varphi = \xi - \gamma = \bar\partial\vartheta\psi \\
= -i(\partial\Lambda\partial\psi - \bar\partial\,\partial\Lambda\psi) \\
= i\,\bar\partial\,\partial\Lambda\psi.
\end{aligned}$$

Thus,

$$\xi - \gamma = \frac{i}{2\pi}\,\partial\,\bar\partial f,$$

where f is a C^∞ function on M. But

$$\xi - \gamma = \bar{\xi} - \bar{\gamma} = -\frac{i}{2\pi} \partial \,\bar\partial f$$

$$= \frac{i}{2\pi} \partial \,\bar\partial \bar{f}$$

since $\xi - \gamma$ is a real form. Hence, $\xi - \gamma = (i/2\pi)\partial \,\bar\partial(\tfrac{1}{2})(f + \bar{f})$ and thus we may assume that $\xi - \gamma = (i/2\pi)\partial \,\bar\partial f$ where f is real valued. Finally

$$\gamma = \xi - \frac{i}{2\pi} \partial \,\bar\partial f = \frac{i}{2\pi} \partial \,\bar\partial(\log \hat{a}_j - f).$$

Let $a_j = \hat{a}_j \, e^{-f}$. Then

$$(1) \quad \gamma = \frac{i}{2\pi} \partial \,\bar\partial \log a_j$$

$$(2) \quad \frac{a_k}{a_j} = |f_{jk}|^2. \qquad \text{Q.E.D.}$$

REMARK.   Perhaps we should explain this proof a little more clearly. We claim:

PROPOSITION 7.1.   If

$$\psi = \frac{1}{2\pi i} \sum \psi_{\alpha\bar\beta} \, dz^\alpha \wedge dz^\beta$$

and if $[\psi] = 0$ (that is, $\psi = d\varphi$), then there is a C^∞ function f such that $\psi = \partial \,\bar\partial f$ when M is a Kähler manifold.

Proof.   Let $\Psi = \{\psi \mid \psi = d\varphi,\ \psi \text{ of type } (1,1)\}$. Then $\partial \,\bar\partial \mathscr{D} \subseteq \Psi$, where $\mathscr{D}$ is the space of differentiable functions, and $\Psi = Y \oplus \partial \,\bar\partial \, \mathscr{D}$, where

$$Y = \{\eta \mid \eta \in \Psi,\ (\eta, \partial\bar\partial f) = 0,\ \text{for all } f \in \mathscr{D}\}.$$

We note that $(\eta, \partial \,\bar\partial f) = 0$ if and only if $\vartheta\bar\vartheta\,\eta = 0$. We claim that if M is Kähler, then $Y = \{0\}$. For Kähler implies $\tfrac{1}{2}\Delta = \square = \bar\square$ and $\partial\vartheta + \vartheta\partial = 0 = \bar\partial\bar\vartheta + \bar\vartheta\bar\partial$. Thus $\tfrac{1}{4}\Delta^2\eta = \square\bar\square\eta = (\bar\partial\bar\vartheta + \bar\vartheta\bar\partial)(\partial\vartheta + \vartheta\partial)\eta$ for $\eta \in Y$. Since η is of type $(1, 1)$ and $\eta = d\varphi$, $0 = d\eta = \partial\eta = \bar\partial\eta$. Thus,

$$\tfrac{1}{4}\Delta^2\eta = (\bar\partial\vartheta\partial\bar\vartheta + \vartheta\partial\bar\partial\bar\vartheta)\eta$$

$$= -\bar\partial\partial\vartheta\bar\vartheta\,\eta + \vartheta\partial\bar\partial\bar\vartheta\,\eta$$

$$= 0.$$

Thus $\Delta^2\eta = 0$ and $(\Delta^2\eta, \eta) = (\Delta\eta, \Delta\eta) = 0$. So $\Delta\eta = 0$ and hence $\delta\eta = 0$. Finally, $(\eta, \eta) = (d\varphi, \eta) = (\varphi, \delta\eta) = 0$ and $\eta = 0$.   Q.E.D.

DEFINITION 7.1.   A complex line bundle F over any compact complex manifold is said to be *positive* if there is a $\gamma = (1/2\pi i) \sum X_{\lambda\bar{\mu}}\, dz^\lambda \wedge dz^{\bar{\mu}}$, $d\gamma = 0$, $\bar{\gamma} = \gamma$, and $[\gamma] = c(F)_{\mathbb{C}}$ such that $X_{\lambda\bar{\mu}}(z)$ is positive definite at every point z of M.

REMARK.   If F over M is positive, then $\omega = i \sum X_{\lambda\bar{\mu}}\, dz^\lambda \wedge dz^{\bar{\mu}}$ is a Kähler form. Hence M is a Kähler manifold. Rewording Theorem 7.1 gives:

THEOREM 7.5.   If $F - K$ is positive, then $H^q(M, \mathcal{O}(F)) = 0$ for $q \geq 1$.

THEOREM 7.6.   If $-F$ is positive, then $H^q(M, \mathcal{O}(F)) = 0$ for $q \leq n - 1$.

Proof.   Serre duality gives $H^q(M, \Omega^p(F)) \cong H^{n-q}(M, \Omega^{n-p}(-F))$ where $\dim M = n$. Notice that $\Omega^n \cong \mathcal{O}(K)$ and let $p = 0$. Then

$$
\begin{aligned}
H^q(M, \Omega^0(F)) &\cong H^{n-q}(M, \Omega^n(-F)) \\
&\cong H^{n-q}(M, \mathcal{O}(K - F)) \\
&= 0 \quad \text{for} \quad n - q \geq 1
\end{aligned}
$$

if $K - F - K = -F$ is positive.   Q.E.D.

We also have:

THEOREM 7.7.   If F is "sufficiently" positive, then $H^q(M, \Omega^p(F)) = 0$ (where F is a line bundle) for $q \geq 1$.

Proof.   Again we use

$$
\begin{aligned}
H^q(M, \Omega^p(F)) &\cong \mathcal{H}^{p,\,q}(F) \\
&= \{\varphi \mid \Box_a \varphi = 0,\ \varphi \text{ of type } (p, q)\}.
\end{aligned}
$$

For $\varphi \in \mathcal{H}^{(p,\,q)}(F)$ we let the reader check the following inequality:

$$
0 \geq \int_M \frac{\omega^n}{n!}\, q \left\{ \sum_{\tau,\,\mu} \sum_{\alpha_i,\,\beta_k} (X^{\bar{\tau}}{}_{\bar{\sigma}} - R_{\bar{\sigma}}{}^{\bar{\tau}}) \varphi_{\alpha_i \cdots \alpha_p \tau \bar{\beta}_2 \cdots \bar{\beta}_q}\, \varphi^{\overline{\bar{\alpha}_1 \cdots \bar{\alpha}_p \sigma \beta_2 \cdots \beta_q}} \right.
$$
$$
\left. + p \sum_{\tau,\,\lambda} \sum_{\mu,\,\sigma} \sum_{\alpha_i,\,\beta_k} R^{\tau}{}_{\lambda\bar{\mu}}{}^{\bar{\sigma}}\, \varphi_{\tau\alpha_2 \cdots \alpha_p \bar{\sigma}\bar{\beta}_2 \cdots \bar{\beta}_q} \cdot \varphi^{\overline{\lambda\alpha_2 \cdots \bar{\alpha}_p \mu \beta_2 \cdots \beta_q}} \right\}.
$$

Thus if $X^{\bar{\tau}}{}_{\bar{\sigma}}$ is sufficiently positive definite, then the integrand is positive for $\varphi \neq 0$ we see $\mathcal{H}^{p,\,q} = 0$.   Q.E.D.

We now proceed to a generalization of Theorem 7.6 due to Nakano (1955). As usual M is a compact Kähler manifold and $F = \{f_{jk}\}$ is a complex line bundle with metric $\{a_j\}$. Remember that $(\vartheta_a \varphi)_j = \{(1/a_j)\vartheta(a_j \varphi_j)\}$, and so forth.

LEMMA 7.2. $(\bar{\partial}\partial_a + \partial_a\bar{\partial})\varphi = X \wedge \varphi$, where $X = -\partial\bar{\partial} \log a_j$ and $\varphi \in \Gamma(A^{p,q}(F))$.

Proof. We have $(\partial_a \varphi)_j = \left\{\dfrac{1}{a_j} \partial(a_j \cdot \varphi_j)\right\}$

$$= \{\partial\varphi_j + \partial \log a_j \wedge \varphi_j\}.$$

Thus,

$$\bar{\partial}(\partial_a \varphi)_j = \{\bar{\partial}\partial\varphi_j + \bar{\partial}\partial \log a_j \wedge \varphi - \partial \log a_j \wedge \bar{\partial}\varphi_j\}.$$

Add $\partial_a\bar{\partial}\varphi_j = \{\partial\bar{\partial}\varphi_j + \partial \log a_j \wedge \bar{\partial}\varphi_j\}$ to get $(\bar{\partial}\partial_a + \partial_a\bar{\partial})\varphi = X \wedge \varphi$.

Q.E.D.

THEOREM 7.8. [Nakano (1955); Calabi and Vesentini (1960)] Let $\varphi \in \Gamma(A^{p,q}(F))$ be such that $\bar{\partial}\varphi = \vartheta_a \varphi = 0$. Then

$$0 \le \sqrt{-1}(X \wedge \Lambda\varphi - \Lambda(X \wedge \varphi), \varphi).$$

Proof. $0 \le \sqrt{-1}\,(\partial_a \varphi, \partial_a \varphi) = (\vartheta\partial_a \varphi, \varphi)$

since $(\bar{\partial}\psi, \psi) = (\varphi, \vartheta_a \psi)$ and $(\vartheta\varphi, \psi) = (\varphi, \partial_a \psi)$. By Proposition 5.4 $-\sqrt{-1}\,\vartheta = \bar{\partial}\Lambda - \Lambda\bar{\partial}$. Hence

$$0 \le (\vartheta\partial_a \varphi, \varphi) = \sqrt{-1}\,(\bar{\partial}\Lambda\partial_a \varphi - \Lambda\bar{\partial}\partial_a \varphi, \varphi) \tag{3}$$

$$= \sqrt{-1}\,(\Lambda\partial_a \varphi, \vartheta_a \varphi) - \sqrt{-1}(\Lambda(\bar{\partial}\partial_a + \partial_a\bar{\partial})\varphi, \varphi)$$

$$= -\sqrt{-1}\,(\Lambda(X \wedge \varphi), \varphi).$$

But we also have

$$0 \le (\vartheta\varphi, \vartheta\varphi) = (\partial_a\vartheta\varphi, \varphi) = \sqrt{-1}\,(\partial_a\bar{\partial}\Lambda\varphi - \partial_a\Lambda\bar{\partial}\varphi, \varphi)$$

$$= \sqrt{-1}\,(\partial_a\bar{\partial}\Lambda\varphi, \varphi)$$

$$= \sqrt{-1}\,(\partial_a\bar{\partial}_{\Lambda\varphi} + \bar{\partial}\partial_a\Lambda\varphi, \varphi)$$

since $\bar{\partial}\varphi = \vartheta_a\varphi = 0$. Thus,

$$\sqrt{-1}\,(X \wedge \Lambda\varphi, \varphi) \ge 0. \tag{4}$$

Now as Equations (3) and (4) to get the theorem.   Q.E.D.

THEOREM 7.9. [Nakano (1955)] If F is negative, then

$$H^q(M, \Omega^p(F)) = 0 \qquad \text{for } p + q \le n - 1$$

when $n = \dim M$.

Proof. By the harmonic theory

$$H^q(M, \Omega^p(F)) \cong \{\varphi \mid \varphi \in \Gamma^{p,q}(F), \bar{\partial}\varphi = \vartheta_a \varphi = 0\} = \mathscr{H}^{p,q}(F),$$

where $\Gamma^{p,q}(F) = \Gamma(A^{p,q}(F))$. By Theorem 7.8 if $\varphi \in \mathscr{H}^{p,q}(F)$

$$0 \le \sqrt{-1}(X \wedge \Lambda\varphi - \Lambda(X \wedge \varphi), \varphi).$$

We let the reader verify the following equations:

$$(X \wedge \varphi)_{\alpha_0 \cdots \alpha_q \bar{\beta}_0 \cdots \bar{\beta}_q} = (-1)^p \sum X_{\alpha_0 \bar{\beta}_0} \varphi_{\alpha_1 \cdots \alpha_p \bar{\beta}_1 \cdots \bar{\beta}_q}$$

$$+ \sum_{i=1}^{p} (-1)^i X_{\alpha_i \bar{\beta}_0} \varphi_{\alpha_0 \cdots \hat{\alpha}_i \cdots \bar{\beta}_1 \cdots}$$

$$+ \sum_{k=1}^{q} (-1)^k X_{\alpha_0 \bar{\beta}_k} \varphi_{\alpha_1 \cdots \alpha_p \bar{\beta}_0 \cdots \hat{\bar{\beta}}_k \cdots \bar{\beta}_q}$$

$$+ \sum_{i=1, k=1}^{p,q} (-1)^{i+k} X_{\alpha_i \bar{\beta}_k} \varphi_{\alpha_0 \cdots \hat{\alpha}_i \cdots \hat{\bar{\beta}}_k \cdots \bar{\beta}_q}$$

$$-\sqrt{-1}\,\Lambda(X \wedge \varphi)_{\alpha_1 \cdots \alpha_p \bar{\beta}_1 \cdots \bar{\beta}_q} = (-1)^p \sum_{\mu,\lambda} g^{\bar{\mu}\lambda}(X \wedge \varphi)_{\lambda\alpha_1 \cdots \bar{\mu}\bar{\beta}_1}$$

$$= \sum_{\mu,\lambda} g^{\bar{\mu}\lambda} X_{\lambda\bar{\mu}} \varphi_{\alpha_1 \cdots \alpha_p \bar{\beta}_1 \cdots \bar{\beta}_q}$$

$$+ \sum_{\mu,\lambda} \sum_{i=1}^{p} (-1)^i g^{\bar{\mu}\lambda} X_{\alpha_i \bar{\mu}} \varphi_{\lambda\alpha_1 \cdots \hat{\alpha}_i \cdots \bar{\beta}_1 \bar{\beta}_2 \cdots}$$

$$+ \sum_{\mu,\lambda} \sum_{k=1}^{q} (-1)^k g^{\bar{\mu}\lambda} X_{\lambda\bar{\beta}_k} \varphi_{\alpha_1 \cdots \bar{\mu}\bar{\beta}_1 \cdots \hat{\bar{\beta}}_k \cdots}$$

$$+ \sum_{\mu,\lambda} \sum_{i=1, k=1}^{p,q} (-1)^{i+k} X_{\alpha_i \bar{\beta}_k} g^{\bar{\mu}\lambda} \varphi_{\lambda \cdots \bar{\mu} \cdots}.$$

Thus,

$$0 \le \int_M \frac{\omega^n}{n!} \frac{1}{p!q!} \{ \sum g^{\bar{\mu}\lambda} X_{\lambda\bar{\mu}} \varphi_{\alpha_1 \cdots \alpha_p \bar{\beta}_1 \cdots \bar{\beta}_q} \varphi^{\overline{\alpha_1 \cdots \alpha_p \bar{\beta}_1 \cdots \bar{\beta}_q}}$$

$$- p \sum g^{\bar{\mu}\lambda} X_{\nu\bar{\mu}} \varphi_{\lambda\alpha_2 \cdots \alpha_p \cdots \alpha_p \bar{\beta}_1 \cdots \bar{\beta}_q} \varphi^{\overline{\nu\alpha_2 \cdots \alpha_p \bar{\beta}_1 \cdots \bar{\beta}_q}}$$

$$- q \sum g^{\bar{\mu}\lambda} X_{\lambda\bar{\nu}} \varphi_{\alpha_1 \cdots \alpha_p \bar{\mu}\bar{\beta}_2 \cdots \bar{\beta}_q} \varphi^{\overline{\alpha_1 \cdots \alpha_p \nu\beta_2 \cdots \beta_q}} \}.$$

Since F is negative, $-X_{\lambda\bar{\mu}}$ is positive definite at each point. (We should use Theorem 7.4 here; that is, we choose a_j so that $-X_{\lambda\bar{\mu}}$ is positive definite.) Now

$$\omega = \sum - X_{\lambda\bar{\mu}}\, dz^\lambda \wedge dz^{\bar{\mu}} = \partial\, \bar{\partial} \log a_j$$

and ω satisfies

$$d\omega = d(\sum - X_{\lambda\bar{\mu}} \, dz^\lambda \wedge dz^{\bar{\mu}}) = 0.$$

Thus we may use $-X_{\lambda\bar{\mu}}$ as a Kähler metric on M. Hence, assume $g_{\alpha\bar{\beta}} = -X_{\alpha\bar{\beta}}$. Then

$$\sum g^{\bar{\mu}\lambda} X_{\nu\bar{\mu}} = -\delta_\nu^\lambda$$

and

$$\sum \delta_\lambda^\lambda = n.$$

Finally

$$0 \leq \int_M \frac{\omega^n}{n!} \frac{1}{p!q!} \{(-n)\sum + p\sum + q\sum\},$$

$$0 \leq \int_M \frac{\omega^n}{n!} \frac{1}{p!q!} (-n + p + q) \sum \varphi_{A_p B_q} \varphi^{\overline{A_p B_q}}.$$

So $0 \leq (-n + p + q)(\varphi, \varphi)$. But $p + q < n$ and we see that φ must be 0.
$$\text{Q.E.D.}$$

8. Hodge Manifolds

Recall that by de Rham's theorem a Kähler form on a manifold M determines an element of $H^2(M, \mathbb{C})$. We also have the image of the canonical map

$$H^2(M, \mathbb{Z}) \longrightarrow H^2(M, \mathbb{C})$$

which we denote $c \to c_{\mathbb{C}}$.

DEFINITION 8.1. $\sum g_{\alpha\bar{\beta}} \, dz^\alpha \, dz^\beta$ is a *Hodge metric* on M if $[\omega] = c_{\mathbb{C}}$ for some $c \in H^2(M, \mathbb{Z})$ where $\omega = i \sum g_{\alpha\bar{\beta}} \, dz^\alpha \wedge dz^\beta$. If M has a Hodge metric, then we say that M is a *Hodge manifold*.

THEOREM 8.1. M is a Hodge manifold if and only if there exists a positive line bundle $F \in H^1(M, \mathcal{O}^*)$.

Proof. Suppose $F \in H^1(M, \mathcal{O}^*)$ is positive. Then, by Theorem 7.4, $c(F) \in H^2(M, \mathbb{Z})$ is cohomologous to $(1/2\pi i) \, X$ where $X = -\partial \bar{\partial} \log a_j = \sum X_{\lambda\bar{\mu}} \, dz^\lambda \wedge dz^{\bar{\mu}}$ and $(X_{\lambda\bar{\mu}})$ is positive definite. Thus $g_{\lambda\bar{\mu}} = (1/2\pi) X_{\lambda\bar{\mu}}$ defines a Hodge metric on M.

Next we assume M has a Hodge metric, that is, $\omega = i \sum g_{\alpha\bar{\beta}} \, dz^\alpha \wedge dz^\beta$ with ω cohomologous to $c_{\mathbb{C}}$ for $c \in H^2(M, \mathbb{Z})$, and $(g_{\alpha\bar{\beta}})$ positive definite. It suffices to show that there is a line bundle F such that $c(F) = c$; because then

$$c(-F) \sim \frac{1}{2\pi i} \sum X_{\alpha\bar{\beta}}\, dz^{\alpha} \wedge dz^{\bar{\beta}},$$

with $X_{\alpha\bar{\beta}} = 2\pi\, g_{\alpha\bar{\beta}}$, and hence $-F$ is positive. Recall the exact sequence

$$\cdots \longrightarrow H^1(M, \mathcal{O}^*) \xrightarrow{\; c \;} H^2(M, \mathbb{Z}) \xrightarrow{\; \mu \;} H^2(M, \mathcal{O}) \longrightarrow \cdots$$
$$F \longrightarrow c(F).$$

Thus, it suffices to show $\mu c = 0$. Let c be defined by the 2-cocycle $c = \{c_{ijk}\}$. The proof involves chasing through the de Rham and Dolbeault isomorphisms. Consider the following diagram:

$$
\begin{array}{c}
\nearrow\; H^2(M, \mathbb{C}) \cong \Gamma(dA^1)/d\Gamma(A^1) \\
c \in H^2(M, \mathbb{Z}) \qquad\qquad c_{\mathbb{C}} \longleftrightarrow \psi \qquad\qquad (1) \\
\searrow^{\mu}\; H^2(M, \mathcal{O}) \cong \Gamma(\bar{\partial}A^{0,1})/\bar{\partial}\Gamma(A^{0,1}) \\
\mu c \longleftrightarrow \varphi^{(0,2)}.
\end{array}
$$

As in the argument of Theorem 7.2, we can find differentiable functions λ_{ij} such that $c_{ijk} = \delta(\lambda)_{ijk} = \lambda_{jk} + \lambda_{ki} + \lambda_{ij}$. Then we can find differentiable 1-forms ψ_j such that $d\lambda_{jk} = \psi_k - \psi_j$. Then ψ in Diagram (1) is obtained by $\psi - d\psi_k = d\psi_j$. For the Dolbeault isomorphism, $\bar{\partial}\lambda_{jk} = \varphi_k - \varphi_j$, where the φ_j are $(0, 1)$-forms. Then $\varphi = \bar{\partial}\varphi_k$. We can split up $\psi_j = \psi_j^{(1,0)} + \psi_j^{(1,0)}$ into forms of type $(1, 0)$ and of type $(0, 1)$. We know that $d = \partial + \bar{\partial}$ so we compute

$$\partial\lambda_{jk} = \psi_k^{(1,0)} - \psi_j^{(1,0)},$$
$$\bar{\partial}\lambda_{jk} = \psi_k^{(0,1)} - \psi_j^{(0,1)}.$$

Thus we may assume that $\varphi_k = \psi_k^{(0,1)}$. Then $\varphi = \bar{\partial}\psi_j^{(0,1)} = \psi^{(0,2)}$ [the $(0, 2)$ part of ψ]. Thus, if $c_{\mathbb{C}} \leftrightarrow \psi$, $\mu c \leftrightarrow \psi^{(0,2)}$. Now we have assumed $c_{\mathbb{C}} \sim \omega$ which is of type $(1, 1)$. Thus $\psi = \omega^{(1,1)} + d\eta$, with $\eta = \eta^{(1,0)} + \eta^{(0,1)}$. Thus $\psi^{(0,2)} = \bar{\partial}\eta^{(0,1)}$ which means $\mu c = 0$.   Q.E.D.

With the obvious definition of elements of type $(1, 1)$ in $H^2(M, \mathbb{Z})$ we have:

COROLLARY.   Let M be a compact complex manifold. Then the image of the map $H^1(M, \mathcal{O}^*) \xrightarrow{c} H^2(M, \mathbb{Z})$ is the set of elements of type $(1, 1)$.

We now give the proof of the main theorem of this chapter which can be considered as a generalization of the fact that every compact Riemann surface is algebraic.

THEOREM 8.2.   [Kodaira (1954)] Every Hodge manifold is algebraic (that is it is a submanifold of some $\mathbb{P}^N$).

We first outline the idea. We know there is a positive line bundle $E \in H^1(M, \mathcal{O}^*)$. Let $F = mE$ where m is a large positive integer. Let

$$\dim H^o(M, \mathcal{O}(F)) = N + 1$$

choose a basis $\{\beta_o, \ldots, \beta_N\}$ for $H^o(M, \mathcal{O}(F))$, and let F be defined by the 1-cocycle $\{f_{jk}\}$ with respect to some covering $\{U_j\}$ of M (remembering that the f_{jk} are never zero). By definition

$$\beta_v = \{\beta_{vj}(z)\}, \; \beta_{vj}(z) = f_{jk}(z) \cdot \beta_{vk}(z),$$

where the $\beta_{vj}(z)$ are holomorphic on U_j. Consider the candidate for a map $\Phi : M \to \mathbb{P}^N$ given by

$$\Phi(z) = (\beta_{oj}(z), \ldots, \beta_{Nj}(z)) \qquad \text{for } z \in U_j.$$

It is easy to see this is well defined as a point of $\mathbb{P}^N$ if for every $z \in M$ there is an index v such that $\beta_v(z) \neq 0$. We want Φ to be an embedding. To prove this it suffices to prove:

(1)   Given $z \in M$, at least one $\beta_v(z) \neq 0$, that is, there is a $\varphi \in H^o(M, \mathcal{O}(F))$, $\varphi = \sum c_v \beta_v$ such that $\varphi(z) \neq 0$. [Then (1) implies that Φ is well defined and holomorphic on M.]

(2) Φ is injective, that is, for any pair of points $p, q \in M$, there is $\varphi \in H^o(M, \mathcal{O}(F))$ such that $\varphi(p) \neq 0$, $\varphi(q) = 0$. [In fact, this also implies (1).]

(3) Φ is biholomorphic, that is, for each point p there exist $n \, (= \dim M)$ elements $\varphi_1, \ldots, \varphi_n \in H^o(M, \mathcal{O}(F))$ such that

$$\det \left(\frac{\partial \varphi_{\alpha j}(z_j)}{\partial z_j^\beta} \right) \neq 0 \qquad \text{at } z_j(p),$$

where $p \in U_j$.

We first prove (2). Let $\mathscr{S} = \mathcal{O}(F - p - q)$ be the subsheaf of $\mathcal{O}(F)$ consisting of germs of holomorphic sections of F which are zero at p and q. Let us investigate the stalks of $\mathscr{S}$. Clearly,

$$\mathscr{S}_z = \mathcal{O}(F)_z, \qquad\qquad \text{if } z \neq p, z \neq q$$
$$\mathscr{S}_p = \{\varphi \in \mathcal{O}(F)_p \, | \, \varphi_j(p) = 0, \; \text{if } p \in U_j\}$$

and similarly for q. We have the exact sequence

$$0 \longrightarrow \mathscr{S} \longrightarrow \mathcal{O}(F) \longrightarrow \mathscr{S}'' \longrightarrow 0, \qquad\qquad (2)$$

where $\mathscr{S}'' = \mathcal{O}(F)/\mathscr{S}$ is the quotient sheaf. Then $\mathscr{S}''_z = 0$ except at p or q. Clearly $\mathscr{S}''_p \cong \mathbb{C}$, $\mathscr{S}''_q \cong \mathbb{C}$ and the isomorphism depends on the choice of local coordinates around p and q. This shows that $H^0(M, \mathscr{S}'') = \mathbb{C} \oplus \mathbb{C}$. The exact cohomology sequence of (2) is

$$0 \longrightarrow H^0(M, \mathscr{S}) \longrightarrow H^0(M, \mathcal{O}(F)) \longrightarrow \mathbb{C} \oplus \mathbb{C} \longrightarrow H^1(M, \mathscr{S}) \longrightarrow \cdots$$
$$\varphi \longrightarrow (\varphi_j(p), \varphi_k(q)).$$

We sometimes use the suggestive notation $\mathscr{S} = \mathcal{O}(F - p - q)$. To prove (2) it is sufficient to prove:

PROPOSITION 8.1.   If M is compact and Kähler and $F \in H^1(M, \mathcal{O}^*)$ is "sufficiently positive," then

$$H^1(M, \mathcal{O}(F - p - q)) = 0.$$

Proof. The proof makes use of the quadric transformations Q_p, Q_q. Let $\tilde{M} = Q_p Q_q(M)$ and let P be the holomorphic map $P : \tilde{M} \to M$ of $\tilde{M}$ onto M such that $C = P^{-1}(p)$ and $D = P^{-1}(q)$ are isomorphic to $\mathbb{P}^{n-1}$ with dim $M = n$, and P is a biholomorphic map on $\tilde{M} - C - D$, $P : \tilde{M} - C - D \to M - p - q$. Let $\mathscr{S} = \mathcal{O}(F - p - q)$, $\tilde{\mathscr{S}} = \mathcal{O}(\tilde{F} - C - D)$, where $\tilde{F}$ is the holomorphic line bundle on $\tilde{M}$ induced by P and $\tilde{\mathscr{S}}$ is the sheaf of germs of holomorphic sections of $\tilde{F}$ which vanish on C and D. Let $\mathscr{U} = \{U_j\}$ be a covering of M. Then $\tilde{\mathscr{U}} = \{\tilde{U}_j\}$, $\tilde{U}_j = P^{-1}(U_j)$ is a covering of $\tilde{M}$. We recall that

$$H^q(M, \mathscr{S}) = \lim_{U} H^q(\mathscr{U}, \mathscr{S})$$

and for $q = 1$, the map $H^1(\mathscr{U}, \mathscr{S}) \to H^1(M, \mathscr{S})$ is injective. We prove:

LEMMA 8.1.   If

$$H^1(\tilde{M}, \mathcal{O}(\tilde{F} - C - D)) = 0,$$

then

$$H^1(M, \mathcal{O}(F - p - q)) = 0.$$

Proof. It suffices to show $H^1(\mathscr{U}, \mathscr{S}) = 0$ for all coverings $\mathscr{U}$. Take a 1-cocycle $\varphi = \{\varphi_{ij}\} \in H^1(\mathscr{U}, \mathscr{S})$, $\varphi_{ij} \in \Gamma(U_i \cap U_j, \mathscr{S})$, where φ_{ij} is a holomorphic section of F over $U_i \cap U_j$ such that $\varphi_{ij}(p) = 0$, $\varphi_{ij}(q) = 0$ if $p, q \in U_i \cap U_j$. P induces

$$\tilde{\varphi}_{ij} = P^* \varphi_{ij} = \varphi_{ij} \circ P \in \Gamma(\tilde{U}_i \cap \tilde{U}_j, \mathcal{O}(F)),$$

where $\tilde{\varphi}_{ij}$ vanishes on C and D if $C \subseteq \tilde{U}_j \cap \tilde{U}_j$, $D \subseteq \tilde{U}_i \cap \tilde{U}_j$. Thus $\{\tilde{\varphi}_{ij}\}$ represents an element of $H^1(\tilde{\mathscr{U}}, \tilde{\mathscr{S}}) \subseteq H^1(\tilde{M}, \tilde{\mathscr{S}}) = 0$. Hence $\tilde{\varphi}_{ij} \sim 0$, that is, $\tilde{\varphi}_{ij} = \psi_j - \psi_i$ where each $\psi_i \in \Gamma(\tilde{U}_i, \mathcal{O}(F))$ and vanishes on C and D. If $U_i \subseteq M - p - q$, then $P : \tilde{U}_i \to U_i$ is biholomorphic. In this case there is $\varphi_i \in \Gamma(U_i, \mathcal{O}(F))$ such that $\psi_i = P^*(\varphi_i)$. If, for instance, $p \in U_i$, then

$P : \tilde{U}_i - C \to U_i - p$ is biholomorphic. Hence there is $\varphi_i \in \Gamma(U_i - p, \mathcal{O}(F))$ such that $P^*(\varphi_i) = \psi_i$ on $\tilde{U} - C$. We can always assume F is trivial over U_i, so $\Gamma(U_i, \mathcal{O}(F)) \cong \Gamma(U_i, \mathcal{O})$. Thus we consider φ_i as a holomorphic function on $U_i - p$. By Hartog's theorem φ_i can be extended to all of U_i. Then $P^*\varphi_i$ is defined on all of $\tilde{U}_i$ and must equal ψ_i (by continuity, or the identity theorem). Thus $\varphi_i(p) = 0$. Hence we have found $\varphi_i \in \Gamma(U_i, \mathcal{O}(F - p))$ such that $\psi_i = P^*\varphi_i$. We have proved that there is a o-cochain $\{\varphi_i\}$, $\varphi_i \in \Gamma(U_i, \mathcal{S})$ such that $\psi_i = P^*\varphi_i$, and thus $P^*\varphi_{ij} = P^*\varphi_i - P^*\varphi_j$. But P is surjective, so $\varphi_{ij} = \varphi_i - \varphi_j$. Thus $\{\varphi_{ij}\} \sim 0$, and $H^1(\mathcal{U}, \mathcal{S}) = 0$.   Q.E.D.

REMARK.   Relations between $H^q(M, \mathcal{S})$ and $H^q(\tilde{M}, \tilde{\mathcal{S}})$ are not easy to see.

To prove Proposition 8.1 it now suffices to prove $H^1(\tilde{M}, \mathcal{O}(\tilde{F} - C - D) = 0$. Let $[C]$ and $[D]$ be the corresponding bundles of the divisors C and D. Then we must show that $H^1(\tilde{M}, \mathcal{O}(\tilde{F} - [C] - [D])) = 0$. To prove this it suffices to show that $\tilde{F} - [C] - [D] - K(\tilde{M})$ is positive, and then quote the vanishing theorem.

We want to show that

$$\tilde{F} - [C] - [D] - K(\tilde{M}) > 0$$

if m is sufficiently large, where $F = mE$, and $K(\tilde{M})$ is the canonical bundle of $\tilde{M}$. Therefore we would like to compute $c([C])$ and $c([D])$. First we find a 1-cocycle on $\tilde{M}$ representing $[C]$. Let z be a coordinate chart map centered at $p \in M$, and let $U = \{z | |z| < 2\varepsilon\}$. Let $P : \tilde{M} \to M$. Let us describe the normal bundle W of C in $\tilde{M}$. Let

$$V_\lambda = \{u \in \mathbb{P}^{n-1} \,|\, u = (u_1, \cdots, u_n), u_\lambda \neq 0\},$$

where $u_1, \cdots, u_n$ are homogeneous coordinates for $\mathbb{P}^{n-1}$. Then

$$\mathbb{P}^{n-1} = \bigcup_{\lambda=1}^{n} V_\lambda,$$

and

$$W = \bigcup_{\lambda=1}^{n} (V_\lambda \times \mathbb{C}),$$

where we identify (u, w_λ) and (v, w_μ) if and only if

$$u = v, \qquad \frac{w_\lambda}{u_\lambda} = \frac{w_\mu}{u_\mu}. \tag{3}$$

We could define $P : W \to U$ by

$$P(u, w_\lambda) = \frac{w_\lambda}{u_\lambda}(u_1, \cdots, u_n) \tag{4}$$

$$= \left(\frac{w_\lambda u_1}{u_\lambda}, \cdots, w_\lambda, \cdots, \frac{w_\lambda u_n}{u_\lambda} \right).$$

Then $\mathbb{P}^{n-1} = \cup_\lambda (V_\lambda \times \{0\}) \subseteq W$ and we can identify $C = Q_p(p)$ with $\mathbb{P}^{n-1}$ in W. Thus we consider a small neighborhood $P^{-1}(U) = \tilde{U}$ of $\mathbb{P}^{n-1}$ in W as a small neighborhood of C in $\tilde{M}$. On each $V_\lambda \times \mathbb{C}$, C is defined by $w_\lambda = 0$. Let $\tilde{U}_\lambda = \tilde{U} \cap (V_\lambda \times \mathbb{C})$ for $\lambda = 1, \cdots, n$, and let $\tilde{U}_0 \subseteq M - C$ be such that

$$Q_p Q_q(M) = \tilde{M} = \tilde{U}_0 \cup \tilde{U}_1 \cup \cdots \cup \tilde{U}_n.$$

We set $w_0 = 1$ on $\tilde{U}_0$. Then the line bundle $[C]$ is given by the 1-cocycle

$$g_{\lambda\nu} = \frac{w_\lambda}{w_\nu} \qquad \text{on } \tilde{U}_\lambda \cap \tilde{U}_\nu. \tag{5}$$

Then (5) implies $g_{\lambda 0} = w_\lambda$.

Recall that, in general, if F is defined by $\{F_{jk}\}$ and if $a_j |f_{jk}|^2 = a_k$ for positive C^∞ functions $\{a_\ell\}$, then

$$c(F) \sim \frac{i}{2\pi} \partial \bar{\partial} \log a_j. \tag{6}$$

We want to find such C^∞ functions for C. We make use of a C^∞ function α on M with the properties

(1) $\alpha(z) = |z|^2$ for $z \in U$, $|z| < \varepsilon$
(2) $\alpha(z) = 1$ for $z \in M - U$.

We define

$$A_0(w) = \alpha(P(w)), \, w \in \tilde{U}_0$$

$$A_\lambda(w) = \frac{\alpha(z)}{|w_\lambda|^2}, \, w \in \tilde{U}_\lambda.$$

Notice, on $C \cap \tilde{U}_\lambda$ local coordinates are $(\lambda \neq 0)$

$$\left(\frac{u_1}{u_\lambda}, \cdots, \frac{u_{\lambda-1}}{u_\lambda}, \frac{u_{\lambda+1}}{u_\lambda}, \cdots, \frac{u_n}{u_\lambda} \right).$$

Thus,

$$A_\lambda(w) = \frac{|z|^2}{|w_\lambda|^2}$$

$$= \left| \frac{1}{u_\lambda} \cdot u \right|^2$$

$$= 1 + \sum_{\nu \neq \lambda} \left| \frac{u_\nu}{u_\lambda} \right|^2$$

so the definition has meaning, and

$$A_\lambda > 0 \text{ on } \tilde{U}_\lambda \qquad \lambda = 0, \cdots, n.$$

The A_λ satisfies

$$A_\lambda |g_{\lambda\nu}|^2 = A_\nu$$

so

$$c([C]) \sim \frac{i}{2\pi}\, \partial\,\bar\partial \log A_\lambda. \tag{7}$$

We notice that

$$\partial\,\bar\partial \log A_\lambda = \partial\,\bar\partial \log\!\left(1 + \sum_{\nu\neq\lambda}\left|\frac{u_\nu}{u_\lambda}\right|^2\right)$$

and this is just the standard Kähler metric on $\mathbb{P}^{n-1} = C$. We also remark that $\partial\,\bar\partial\,\alpha(z)$ is a C^∞ 2-form on M of type $(1, 1)$. Since

$$\partial\,\bar\partial\,\alpha = d(\bar\partial\alpha)$$

$\partial\,\bar\partial\,\alpha$ is cohomologous to zero on M, and the induced form $\Omega = P^*(\partial\bar\partial\alpha)$ is cohomologous to zero on $\tilde M$. We then define

$$\sigma_c = \frac{1}{2\pi i}\,(\partial\,\bar\partial \log A_\lambda + \Omega).$$

Then

$$\sigma_c \sim c(-[C]) \tag{8}$$

by (7), and in a neighborhood of C,

$$2\pi i\sigma_c = \partial\,\bar\partial \log\!\left(1 + \sum_{\nu\neq\lambda}\left|\frac{u_\nu}{u_\lambda}\right|^2\right) + \partial\,\bar\partial(|z|^2).$$

Recall that

$$z = (z_1, \cdots, z_n) = \frac{w_\lambda}{u_\lambda}u = (\cdots, w_\lambda, \cdots).$$

Then

$$\partial\bar\partial\Big(\sum z_\nu\bar z_\nu\Big) = \sum dz_\nu \wedge d\bar z_\nu$$
$$= dw_\lambda \wedge dw_\lambda + \cdots.$$

Hence σ_c is positive definite in a neighborhood of C. We get similar results for D.

Next we want to find a relation between $K(M) = K$ and $K(\tilde M)$. We prove:

PROPOSITION 8.2. $K(\tilde M) = \tilde K + (n-1)\,[C] + (n-1)[D]$, where $\tilde K$ is the bundle over M induced from the canonical bundle K of M.

Proof. Suppose $\tilde M = Q_p(M)$. It is sufficient to prove

$$K(\tilde M) = \tilde K + (n-1)\,[C]. \tag{9}$$

We choose $U \ni p$ as in the previous proof. Then we choose $\{U_j, z_j\}$ coordinate systems on M such that $\{U\} \cup \{U_j\}$ covers M. Let $(z^1, \cdots, z^n)$ be a coordinate system on M. Then the canonical bundle K of M is defined by the 1-cocycle $\{J_{jk}\}$ where

$$dz_j^1 \wedge \cdots \wedge dz_j^n = J_{jk}^{-1} \, dz_k^1 \wedge \cdots \wedge dz_k^n \qquad \text{on } U_j \cap U_k,$$

$$dz^1 \wedge \cdots \wedge dz^n = J_{ok}^{-1} \, dz_k^1 \wedge \cdots \wedge dz_k^n \qquad \text{on } U \cap U_k.$$

On $\tilde{M}$ we may use $\{\tilde{U}_\lambda, \tilde{U}_j\}$ as a coordinate covering where $\tilde{U}_j = P^{-1}(U_j) = U_j$ and $\tilde{U} = P^{-1}(U) = \cup_{\lambda=1}^n \tilde{U}_\lambda$, using the notation of the previous theorem. On $\tilde{U}_\lambda$ we have the local coordinate system

$$\left(\frac{u_1}{u_\lambda}, \cdots, w_\lambda, \cdots, \frac{u_n}{u_\lambda} \right)$$

and

$$z^\alpha = \frac{w_\lambda u_\alpha}{u_\lambda}, \qquad \text{if } \alpha \neq \lambda$$

$$z^\lambda = w^\lambda, \qquad \text{if } \alpha = \lambda.$$

Computing, we get

$$P^*(dz^1 \wedge \cdots \wedge dz^n) = d\left(\frac{w_\lambda u_1}{u_\lambda} \right) \wedge \cdots \wedge d\left(\frac{w_\lambda u_{\lambda-1}}{u_\lambda} \right) \wedge dw_\lambda \wedge \cdots \wedge d\left(\frac{w_\lambda u_n}{u_\lambda} \right)$$

$$= w_\lambda^{n-1} \, d\left(\frac{u_1}{u_\lambda} \right) \wedge \cdots \wedge dw_\lambda \wedge \cdots d\left(\frac{u_n}{u_\lambda} \right)$$

since

$$dz^\alpha = d\left(\frac{w_\lambda u_\alpha}{u_\lambda} \right) = w_\lambda \, d\left(\frac{u_\alpha}{u_\lambda} \right) + \left(\frac{u_\alpha}{u_\lambda} \right) dw_\lambda.$$

Let $\{I_{jk}, I_{\lambda v}, I_{\lambda k}\}$ be the Jacobians on $\tilde{M}$ [which are used to define $K(\tilde{M})$]. Then

$$I_{jk}^{-1} = J_{jk}^{-1}$$

$$I_{\lambda k}^{-1} = \frac{1}{w_\lambda^{n-1}} \, J_{ok}^{-1}$$

so

$$I_{\lambda v} = \frac{w_\lambda^{n-1}}{w_v^{n-1}}.$$

This proves (9) since $w_\lambda = 0$ defines c.

We now return to the proof of Proposition 8.1. Recall that E is positive, that is,

$$c(E) \sim \gamma = -i \sum \gamma_{\alpha\bar\beta}\, dz^\alpha \wedge dz^\beta,$$

where $(\gamma_{\alpha\bar\beta})$ is positive definite. For simplicity we write $\gamma > 0$ or $E > 0$. We want to show

$$\tilde F - [C] - [D] - K(\tilde M) = \tilde F - \tilde K - n[C] - n[D] > 0 \qquad (10)$$

for large m. Let $c(K) \sim \kappa$. Then

$$c(\tilde F - \tilde K - n[C] - n[D]) \sim m\tilde\gamma - \tilde\kappa + n\sigma_C + n\sigma_D,$$

where $\tilde\gamma = P^*\gamma$ and $\tilde\kappa = P^*\kappa$. We choose m so large that $m\gamma - \kappa$ is positive definite on M. Then $m\tilde\gamma - \tilde\kappa$ is positive semidefinite on $\tilde M$ and is positive definite on $\tilde M - C - D$. But $\sigma_C > 0$ near C and $\sigma_D > 0$ near D. Then Equation (10) follows. This proves Proposition 8.1, and thus part (1) and (2).

REMARK.   It is an easy compactness argument to see that one can find an integer m such that $H^0(M, mE)$ separates points for all $p, q \in M$, $p \neq q$.

The proof of C is almost the same as the proof of B. We want to show that Φ is biholomorphic at each $p \in M$. Consider $\mathscr{S} = \mathcal{O}(F - 2p)$ which is the sheaf of germs of holomorphic sections of F which vanish at p up to order 2. Again we compute the stalks $\mathscr{S}_z$ and write down the exact sequence

$$0 \longrightarrow \mathscr{S} \longrightarrow Q(F) \longrightarrow \mathscr{S}'' \longrightarrow 0,$$

$$\mathscr{S}_z = \mathcal{O}(F)_z, \qquad z \neq p$$

$$\mathscr{S}_p = \left\{ \varphi \,\middle|\, \varphi = \varphi_j,\, \varphi_j(z) = \sum_{k_1 + \cdots + k_n \geq 2} a_{k_1 \cdots k_n} z_j^{k_1} \cdots z_j^{k_n},\, \varphi \in \mathcal{O}(F)_p \right\}.$$

Then

$$\mathscr{S}''_z = 0, \qquad\qquad\qquad\quad \text{if } z \neq p$$

$$= \left\{ \varphi \,\middle|\, \varphi = a_o + \sum_{\alpha=1}^{n} a_\alpha z^\alpha \right\}, \qquad \text{if } z = p.$$

Thus $H^0(M, \mathscr{S}'') \cong \mathbb{C}^{n+1}$. We write down the exact cohomology sequence

$$0 \longrightarrow H^0(M, \mathscr{S}) \longrightarrow H^0(M, \mathcal{O}(F)) \longrightarrow H^0(M, \mathscr{S}'')$$

$$\overset{\delta}{\longrightarrow} H^1(M, \mathscr{S}) \longrightarrow \cdots.$$

It is easily seen that to prove Φ is biholomorphic at p we need only show $H^1(M, \mathscr{S}) = 0$. To prove this we once again use $\tilde M = Q_p(M)$, $C = Q_p(p)$.

LEMMA 8.2. If

$$H^1(\tilde{M}, \mathcal{O}(\tilde{F} - 2[C])) = 0,$$

then

$$H^1(M, \mathcal{O}(F - 2p)) = 0.$$

Proof. The proof is the same as that of Lemma 8.1. One only has to notice that if φ has a zero of order 2 at p, $P^*\varphi$ has a zero of order 2 (at least) on C and vice versa.

LEMMA 8.3. $H^1(\tilde{M}, \mathcal{O}(\tilde{F} - 2[C])) = 0$ if m is large enough where $F = mE$.

Proof. Using Proposition 8.1. we find

$$\tilde{F} - 2[C] - K(\tilde{M}) = m\tilde{E} - \tilde{K} - (n-1)[C] - 2[C]$$
$$= m\tilde{E} - \tilde{K} - (n+1)[C].$$

Hence

$$c(\tilde{F} - 2[C] - K(\tilde{M})) \sim m\tilde{\gamma} - \tilde{K} + (n+1)\sigma_C > 0$$

if m is large enough.   Q.E.D.

RFMARK.   We again use compactness to see that there is an m which will work for all $p \in M$. This completes the proof of Theorem 8.2.

We now derive some consequences:

THEOREM 8.3.   [Kodaira (1960)] If M is compact Kähler and $H^2(M, \mathcal{O}) = 0$, then M is projective algebraic.

Proof. The exact cohomology sequence of

$$0 \longrightarrow \mathbb{Z} \longrightarrow \mathcal{O} \longrightarrow \mathcal{O}^* \longrightarrow 0$$

yields

$$\cdots \longrightarrow H^1(M, \mathcal{O}^*) \overset{c}{\longrightarrow} H^2(M, \mathbb{Z}) \longrightarrow 0.$$

Thus everything in $H^2(M, \mathbb{Z})$ is the Chern class of some bundle. Let $\{b_1, \cdots, b_m\}$ be a basis for the free part of $H^2(M, \mathbb{Z})$ so that

$$H^2(M, \mathbb{C}) = \mathbb{C}b_1 + \cdots + \mathbb{C}b_m.$$

Each $b_\lambda = c(F_\lambda)$ and hence is cohomologous to a real 2-form of type $(1, 1)$. Let

$$\omega = i \sum g_{\alpha\bar\beta} \, dz^\alpha \wedge dz^\beta$$

be a Kähler form on M. We wish to modify ω to get a Hodge metric on M. Since $\omega \in H^2(M, \mathbb{C})$

$$\omega \sim \sum \rho_\lambda b_\lambda$$

where $\rho_\lambda \in \mathbb{R}$ (ω is real and the b_λ are real). Given ε, we can always find integers $k_\lambda, r \in \mathbb{Z}$ such that

$$\left| \rho_\lambda - \frac{k_\lambda}{r} \right| < \varepsilon \qquad \lambda = 1, \cdots, m.$$

But then for a small enough ε

$$\omega' = \omega - \sum \left(\frac{\rho_\lambda - k_\lambda}{r} \right) \gamma_\lambda$$

defines a Kähler form on M where $\gamma_\lambda \sim b_\lambda$ is a real $(1, 1)$ form. Hence $\tilde\omega = r\omega'$ is also a Kähler form. But

$$\tilde\omega \sim r \sum \rho_\lambda b_\lambda - r \sum \left(\rho_\lambda - \frac{k_\lambda}{r} \right) b_\lambda = \sum k_\lambda b_\lambda \in H^2(M, \mathbb{Z}).$$

Thus $\tilde\omega$ defines a Hodge metric on M, and M is algebraic.   Q.E.D.

Theorem 8.4.   [Kodaira (1954)]   Let M be a compact complex manifold. If the universal covering manifold $\tilde M$ is complex analytically homeomorphic to a bounded domain $\mathscr{B} \subseteq \mathbb{C}^n$, then M is algebraic.

Proof. We make use of the Bergmann metric on $\mathscr{B}$ [see Helgason (1962)]. We have $M = \mathscr{B}/G$ where G, the set of covering transformations of $\mathscr{B}$, is a collection of biholomorphic maps from $\mathscr{B}$ to $\mathscr{B}$. Let $ds^2 = \sum g_{\alpha\bar\beta} \, dz^\alpha \, d\bar z^\beta$ be the Bergmann metric on $\mathscr{B}$. We claim

(1) ds^2 is invariant under G and hence induces a metric $\sum g_{\alpha\bar\beta} \, dz^\alpha \, d\bar z^\beta$ on $M = \mathscr{B}/G$.

(2)   If $\omega = (i/2\pi) \sum g_{\alpha\bar\beta} \, dz^\alpha \wedge d\bar z^\beta$, $\omega \sim c(-K)$; so we have a Hodge metric on M.

This gives the theorem, thus we need only prove (1) and (2).

Let $\mathscr{H}$ be the Hilbert space of all holomorphic functions f on $\mathscr{B}$ which have bounded norm

$$\|f\|^2 = \int_{\mathscr{B}} |f(z)|^2 \, dX,$$

where

$$dX = dx_1 \cdots dx_{2n} \quad \text{and} \quad z_\alpha = x_{2\alpha-1} + ix_{2\alpha}.$$

Let $\{f_\nu\}$ be any orthonormal base of $\mathscr{H}$. Then the Bergmann kernel $K(z, \bar{z})$ is given by

$$K(z, \bar{z}) = \sum_{\nu=1}^{\infty} f_\nu(z)\overline{f_\nu(z)} \qquad [= K(z)].$$

The kernel $K(z)$ is actually independent of the choice of orthonormal basis $\{f_\nu\}$ [see Helgason (1962)]. Then $\sum h_{\alpha\bar{\beta}} \, dz^\alpha \, d\bar{z}^\beta$ is a positive definite Hermitian metric where

$$h_{\alpha\bar{\beta}}(z) = \frac{\partial^2 \log K(z)}{\partial z_\alpha \, \partial \bar{z}_\beta}.$$

Let $\gamma : \mathscr{B} \to \mathscr{B}$ be a biholomorphic map, $\gamma(z) = z'$.

LEMMA 8.4.

$$K(z) = \left| \det \frac{\partial(z'_1, \cdots, z'_n)}{\partial(z_1, \cdots, z_n)} \right|^2 K(z').$$

Proof.

$$K(z) = \sum_{\nu=1}^{\infty} f_\nu(z)\bar{f}_\nu(z)$$

and

$$\int_{\mathscr{B}} f_\nu(z')\overline{f_\lambda(z')} \, dX' = \int_{\mathscr{B}} f_\nu(z)\overline{f_\lambda(z)} \, dX = \delta_{\nu\lambda}.$$

Let

$$F_\nu(z) = f_\nu(z') \det\left(\frac{\partial(z')}{\partial(z)} \right)$$

and notice

$$dX' = \left| \det\left(\frac{\partial(z')}{\partial(z)} \right) \right|^2 dX.$$

Thus,

$$\int_{\mathscr{B}} f_\nu \bar{f}_\lambda \, dX' = \int_{\mathscr{B}} F_\nu \bar{F}_\lambda \, dX = \delta_{\nu\lambda}$$

and $\{F\}$ gives a new base. Hence,

$$K(z) = \sum_{\nu=1}^{\infty} F_\nu(z)\bar{F}_\nu(z) = \left| \det \frac{\partial(z')}{\partial(z)} \right|^2 \sum f_\nu(z')\bar{f}_\nu(z')$$

$$= \left| \det \frac{\partial(z')}{\partial(z)} \right|^2 K(z'). \qquad \text{Q.E.D.}$$

Since G is a group of biholomorphic maps this proves (1).

Now let K be the canonical bundle of M. Let $\pi : B \to B/G = M$. Let U_j be an open set in M on which a local inverse of π is defined, and choose one $\mu_j = \pi^{-1}$ to use as a coordinate chart for U_j ($\mu_j(p) \in \mathbb{C}^n$ if $p \in U_j$). Suppose $p \in U_j \cap U_k$. Then there is $\gamma_{jk} \in G$ such that $\mu_k(p) = \gamma_{jk}(\mu_j(p))$. The canonical bundle K on M is defined by the 1-cocycle

$$f_{jk} = \det \left(\frac{\partial(z_k)}{\partial(z_j)} \right)$$

and we have

$$K(z_j) = |f_{jk}|^2 \, K(z_k)$$

by the lemma. Recall that if we have positive C^∞ functions a_j on U_j such that $a_j |f_{jk}|^2 = a_k$, then

$$c(-K) = -\frac{i}{2\pi} \, \partial \, \bar\partial \log a_j.$$

Therefore, let $a_j = K^{-1}(z_j)$. Then

$$c(-K) = \frac{i}{2\pi} \, \partial\bar\partial \log K(z_j) = \frac{i}{2\pi} \sum g_{\alpha\bar\beta} \, dz_j^\alpha \wedge dz_j^{\bar\beta}.$$

This proves the theorem.

REMARK.   There is much interest in nonalgebraic Kähler manifolds. Kähler manifolds give examples of the minimal surfaces of differential geometry.

[4]

Applications of Elliptic Partial Differential Equations to Deformations

I.   Infinitesimal Deformations

We want to study analytic families of compact, complex manifolds. Informally, we are only interested in *small deformations*. We may as well assume our base space $B_r = \{t \mid |t| < r, t \in \mathbb{C}^m\}$ is an open disk around the origin of $\mathbb{C}^m$. We want a manifold $\mathcal{M}$ and a holomorphic map $\bar{\omega} : \mathcal{M} \to B_r$ with maximal rank so that $\bar{\omega}$ is proper and each fibre $M_t = \bar{\omega}^{-1}(t)$ has the structure of a complex manifold which varies analytically with t. We want a covering $\{\mathcal{U}_j\}$ of $\mathcal{M}$ so that

$$\mathcal{U}_j = \{(\zeta_j, t) \mid |\zeta_j| < 1, |t| < r\}$$

$$\zeta_j = (\zeta_j^1, \cdots, \zeta_j^n), \; \bar{\omega}(\zeta_j, t) = t,$$

and

$$\zeta_j^\alpha = f_{jk}^\alpha(\zeta_j, t) \text{ on } \mathcal{U}_j \cap \mathcal{U}_k,$$

where f_{jk} is holomorphic in ζ_j and t. We notice that under these circumstances M_t is diffeomorphic to M_0, and in fact, $\mathcal{M}$ is diffeomorphic to $X \times B_r$, where X is the underlying differentiable manifold of M_0. Thus $\mathcal{U}_j = U_j \times B_r$ where $U_j = \{\zeta_j \mid \zeta_j < 1\}$, and

$$M = \bigcup_j U_j \times B_r.$$

If x is a point of X, $t \in B_r$ we notice that

$$\zeta_j^\alpha = \zeta_j^\alpha(x, t)$$

is a differentiable function of (x, t) and we have

$$\zeta_j^\alpha(x, t) = f_{jk}^\alpha(\zeta_k(x, t), t). \tag{1}$$

Let $M = M_0 = X$ and use the complex coordinates z of M as differentiable coordinates so that

$$\zeta_j^\alpha(x, t) = \zeta_j^\alpha(z, t),$$

where $\zeta_j^\alpha(z, t)$ is a *differentiable* function of z and t. Because $t = 0$, $\zeta_j^\alpha(z, 0)$ is holomorphic in z (otherwise it is only differentiable).

147

DEFINITION 1.1.   Let

$$\left(\frac{\partial}{\partial t}\right) = \sum_{v=1}^{m} c_v \left(\frac{\partial}{\partial t_v}\right)$$

belong to the tangent space to B_r at the origin. We define $(\partial M_t/\partial t)_{t=0}$ to be the cohomology class in $H^1(M, \Theta)$ given by the 1-cocycle

$$\theta_{ik} = \sum_{\alpha=1}^{n} \left.\frac{\partial f_{ik}^{\alpha}(\zeta_k, t)}{\partial t}\right|_{t=0} \left(\frac{\partial}{\partial \zeta_i^{\alpha}}\right).$$

We want to represent $\{\theta_{ik}\}$ with a 1-form by using Dolbeault's theorem. Let T be the holomorphic tangent bundle on M and $\Theta = \mathcal{O}(T)$ be the sheaf of sections. Then we have the resolution

$$0 \longrightarrow \Theta \longrightarrow A^0(T) \overset{\bar{\partial}}{\longrightarrow} A^{0,\,1}(T) \overset{\bar{\partial}}{\longrightarrow} \cdots,$$

where $A^{0,q}(T)$ is the sheaf of C^∞ vector $(0, q)$-forms. Locally, such a thing has the representation

$$\varphi = \sum_{\beta=1}^{n} \varphi^\beta \left(\frac{\partial}{\partial z^\beta}\right),$$

where

$$\varphi^\beta = \frac{1}{q!} \sum \varphi_{\bar{\alpha}_1 \cdots \bar{\alpha}_q}^{\beta} \, d\bar{z}^{\alpha_1} \wedge \cdots \wedge d\bar{z}^{\alpha_q}$$

and

$$\bar{\partial}\varphi = \sum_{\beta=1}^{n} \bar{\partial}\varphi^\beta \left(\frac{\partial}{\partial z^\beta}\right).$$

Let us trace through the Dolbeault isomorphism

$$H^1(M, \Theta) \cong \frac{\Gamma(\bar{\partial} A^0(T))}{\bar{\partial}\Gamma(A^0(T))}.$$

Let $(\partial M_t/\partial t)_{t=0} \leftrightarrow \eta$. Then η is defined as follows: Pick $\xi_i \in \Gamma(U_i, A^0(T))$ such that

$$\theta_{ik} = \xi_k - \xi_i \text{ on } U_k \cap U_k.$$

Then

$$\eta = \bar{\partial}\xi_i = \bar{\partial}\xi_k.$$

PROPOSITION 1.1.

$$\eta = -\sum_{\alpha=1}^{n} \bar{\partial}\left(\left.\frac{\partial \zeta_i^{\alpha}(z, t)}{\partial t}\right|_{t=0}\right)\left(\frac{\partial}{\partial \zeta_i^{\alpha}}\right).$$

Proof.   Let

$$\dot{\zeta}_i^\alpha = \left(\frac{\partial}{\partial t}\,\zeta_i^\alpha(z, t)\right)_{t=0}.$$

Then Equation (1) yields

$$\dot{\zeta}_i^\alpha = \sum_\beta \frac{\partial f_{ik}^\alpha}{\partial \zeta_k^\beta}\,\dot{\zeta}_k^\beta + \left(\frac{\partial f_{ik}^\alpha}{\partial t}\right)_{t=0}$$

Thus,

$$\theta_{ik} = \sum_\alpha \left(\frac{\partial f_{ik}^\alpha}{\partial t}\right)\left(\frac{\partial}{\partial \zeta_i^\alpha}\right)$$

$$= \sum \dot{\zeta}_i^\alpha\left(\frac{\partial}{\partial \zeta_i^\alpha}\right) - \sum \dot{\zeta}_k^\beta\left(\frac{\partial \zeta_i^\alpha}{\partial \zeta_k^\beta}\right)\left(\frac{\partial}{\partial \zeta_i^\alpha}\right)$$

$$= \sum \dot{\zeta}_i^\alpha\left(\frac{\partial}{\partial \zeta_i^\alpha}\right) - \sum \dot{\zeta}_k^\beta\left(\frac{\partial}{\partial \zeta_k^\beta}\right).$$

If we set $\xi_k = -\sum \dot{\zeta}_k^\beta(\partial/\partial\zeta_k^\beta)$, we get $\theta_{ik} = \xi_k - \xi_i$. Therefore, $\eta = \bar\partial\xi_i = -\sum \bar\partial\dot{\zeta}_i^\alpha(\partial/\partial\zeta_i^\alpha)$.   Q.E.D.

We want to define a vector $(0, 1)$-form $\varphi(t)$, $t \in B$ which describes the complex structure of M_t. With respect to the local complex coordinate z on a neighborhood W of M we have the differential operators

$$\partial_\beta = \left(\frac{\partial}{\partial z^\beta}\right), \quad \bar\partial_\beta = \left(\frac{\partial}{\partial \bar z^\beta}\right),$$

and

$$\bar\partial = \sum_\beta d\bar z^\beta\left(\frac{\partial}{\partial \bar z^\beta}\right).$$

Recall that

$$\zeta_j^\alpha(z, t) = f_{jk}^\alpha(\zeta_k(z, t), t),$$

where f_{jk}^α is a holomorphic function of ζ_k. Thus

$$\bar\partial\zeta_j^\alpha(z, t) = \sum_\beta \frac{\partial f_{jk}^\alpha}{\partial \zeta_k^\beta}\,\bar\partial\zeta_k^\beta(z, t),$$

and

$$\partial_\lambda\zeta_j^\alpha(z, t) = \sum_\beta \frac{\partial f_{jk}^\alpha}{\partial \zeta_k^\beta}\,\partial_\lambda\zeta_k^\beta(z, t).$$

At $t = 0$, $(z^1, \cdots, z^n)$ are local complex coordinates on M and $(\zeta_j^1(z, 0), \cdots, \zeta_j^n(z, 0))$ are also local coordinates. Hence

$$\det\left(\frac{\partial \zeta_j^\alpha(z, 0)}{\partial z^\lambda}\right) \neq 0.$$

So, for small enough $|t|$,

$$\det\left(\frac{\partial \zeta_j^\alpha(z, t)}{\partial z^\lambda}\right) \neq 0.$$

Let

$$A_{j\alpha}^\lambda = \left(\frac{\partial \zeta_j^\alpha}{\partial z^\lambda}\right)^{-1}.$$

Consider the local 1-form,

$$\sum_\alpha A_{j\alpha}^\lambda \, \bar{\partial}\zeta_j^\alpha(z, t) = \varphi_j^\lambda(z, t).$$

We claim the form $\varphi_j^\lambda(a, t)$ is well defined independent of ζ_j. For $f_{jk}^\alpha(\zeta_k, t)$ holomorphic in ζ_k implies

$$A_{j\alpha}^\lambda = \sum_\beta A_{k\beta}^\lambda \frac{\partial \zeta_k^\beta}{\partial \zeta_j^\alpha}$$

which yields

$$\varphi_j^\lambda(z, t) = \sum_{\alpha, \beta} \frac{\partial \zeta_j^\alpha}{\partial \zeta_k^\beta} A_{j\alpha}^\lambda \, \bar{\partial}\zeta_k^\beta(z, t)$$

$$= \sum_\beta A_{k\beta}^\lambda \, \bar{\partial}\zeta_k^\beta(z, t).$$

We therefore define $\varphi^\lambda = \varphi_j^\lambda(z, t)$. Then $\varphi^\lambda(z, t)$ is a $(0, 1)$-form independent of ζ_j, but it still depends on the local coordinate z. Let $z = z_W$. Suppose V is another open set in M with local coordinate z_V. Let $\varphi_W^\lambda = \varphi^\lambda(z_W, t)$ and $\varphi_V^\lambda = \varphi^\lambda(z_V, t)$.

    If we set

$$\varphi_W = \sum \varphi_W^\beta \left(\frac{\partial}{\partial z_W^\beta}\right),$$

then

$$\varphi(t) = \sum_\beta \varphi_W^\beta(z_W, t)\left(\frac{\partial}{\partial z_W^\beta}\right)$$

$$= \sum_\beta \varphi_V^\beta(z_V, t)\left(\frac{\partial}{\partial z_V^\beta}\right)$$

is a well-defined, *global* vector $(0, 1)$-form on M. Now by the definition of $A^{\lambda}_{j\alpha}$,

$$\bar{\partial}\zeta^{\alpha}_{j}(z, t) = \sum \varphi^{\lambda}\frac{\partial\zeta^{\alpha}_{j}}{\partial z^{\lambda}}.$$

Therefore

$$\left(\bar{\partial} - \sum_{\lambda}\varphi^{\lambda}\partial_{\lambda}\right)\zeta^{\alpha}_{j}(z, t) = 0. \tag{2}$$

PROPOSITION 1.2.   The complex structure on M_t is determined by $\varphi(t)$. More specifically a differentiable function f defined on any open subset of M is holomorphic with respect to the complex structure of M_t if and only if

$$(\bar{\partial} - \sum \varphi^{\beta}(t)\partial_{\beta})f(z) = 0. \tag{3}$$

Proof. We represent $\varphi^{\beta}(t)$ by

$$\varphi^{\beta}(t) = \sum_{\alpha}\varphi(t)^{\beta}_{\bar{\alpha}}\,d\bar{z}^{\alpha}.$$

If f satisfies (2), then

$$\left(\bar{\partial}_{\alpha} - \sum_{\beta}\varphi(t)^{\beta}_{\bar{\alpha}}\partial_{\beta}\right)f = 0 \qquad \text{for all } \alpha. \tag{4}$$

We use

$$\frac{\partial f}{\partial z^{\beta}} = \sum_{\gamma}\frac{\partial f}{\partial\zeta^{\gamma}_{j}}\frac{\partial\zeta^{\gamma}_{j}}{\partial z^{\beta}} + \sum_{\gamma}\frac{\partial f}{\partial\bar{\zeta}^{\gamma}_{j}}\frac{\partial\bar{\zeta}^{\gamma}_{j}}{\partial z^{\beta}}$$

and Equation (4) to get

$$\sum_{\gamma}\frac{\partial f}{\partial\zeta^{\gamma}_{j}}\left(\bar{\partial}_{\alpha} - \sum_{\beta}\varphi(t)^{\beta}_{\bar{\alpha}}\partial_{\beta}\right)\zeta^{\gamma}_{j} + \sum_{\gamma}\frac{\partial f}{\partial\bar{\zeta}^{\gamma}_{j}}(\bar{\partial}_{\alpha} - \sum\varphi(t)^{\beta}_{\bar{\alpha}}\partial_{\beta})\bar{\zeta}^{\gamma}_{j} = 0.$$

Equation (2) implies that the first term is zero so

$$\sum_{\gamma}\frac{\partial f}{\partial\bar{\zeta}^{\gamma}_{j}}\left(\bar{\partial}_{\alpha} - \sum_{\beta}\varphi(t)^{\beta}_{\bar{\alpha}}\partial_{\beta}\right)\bar{\zeta}^{\gamma}_{j} = 0. \tag{5}$$

Since $\zeta^{\alpha}_{j}(z, 0)$ is holomorphic in z, $\varphi(0) = 0$. Continuity tells us that $\varphi(t)$ is small for small t. Thus,

$$\left(\bar{\partial}_{\alpha} - \sum_{\beta}\varphi(t)^{\beta}_{\bar{\alpha}}\partial_{\beta}\right)\bar{\zeta}^{\gamma}_{j}$$

is invertible for small t. Hence Equation (5) yields

$$\frac{\partial f}{\partial\bar{\zeta}^{\gamma}_{j}} = 0.$$

Thus f is holomorphic in $\zeta_j^\alpha(z, t)$. We can read the argument backwards, so the proof is concluded.   Q.E.D.

We want to introduce a bracket operation in the algebra of vector $0, q)$-forms. Let us recall the Lie bracket of vector fields. If $u = \sum u^\alpha \partial_\alpha$, $w = \sum w^\alpha \partial_\alpha$, then

$$[u, w] = \sum_{\alpha, \beta} (u^\alpha \partial_\alpha w^\beta - w^\alpha \partial_\alpha u^\beta)\partial_\beta.$$

For the generalization, let

$$\varphi^\alpha = \frac{1}{p!} \sum \varphi^\alpha_{\bar\lambda_1 \cdots \bar\lambda_p} \, d\bar{z}^{\lambda_1} \wedge \cdots \wedge d\bar{z}^{\lambda_p}$$

$$\psi^\alpha = \frac{1}{q!} \sum \psi^\alpha_{\bar\mu_1 \cdots \bar\mu_q} \, d\bar{z}^{\mu_1} \wedge \cdots \wedge d\bar{z}^{\mu_q}.$$

DEFINITION 1.2.

$$[\varphi, \psi] = \sum_{\alpha, \beta = 1}^{n} (\varphi^\alpha \wedge \partial_\alpha \psi^\beta - (-1)^{pq}\psi^\alpha \wedge \partial_\alpha \varphi^\beta)\partial_\beta$$

where

$$\partial_\alpha \psi^\beta = \frac{1}{p!} \sum \partial_\alpha \varphi^\beta_{\bar\lambda_1 \cdots \bar\lambda_p} \, d\bar{z}^{\lambda_1} \wedge \cdots \wedge d\bar{z}^{\lambda_p}.$$

PROPOSITION 1.3.

(1) $[\varphi, \psi]$ is bilinear
(2) $[\psi, \varphi] = -(-1)^{pq}[\varphi, \psi]$
(3) $\bar\partial[\varphi, \psi] = [\bar\partial\varphi, \psi] + (-1)^p[\varphi, \bar\partial\psi]$
(4) $(-1)^{pr}[\varphi, [\psi, \tau]] + (-1)^{qp}[\psi, [\tau, \varphi]] + (-1)^{rq}[\tau, [\varphi, \psi]] = 0$,

where φ is a $(0, p)$-form, ψ is $(0, q)$ and, τ is $(0, r)$. (This is the Jacobi identity.)

Proof. Uninteresting; we leave it to the reader. We collect our facts into the following theorem:

THEOREM 1.1. If $\bar\omega : \mathcal{M} \to B_r$ is a complex analytic family of compact complex manifolds, then the complex structure on $M_t = \bar\omega^{-1}(t)$ is represented by a vector $(0, 1)$-form $\varphi(t)$ on M_0 such that

(1) $\bar\partial\varphi(t) - \frac{1}{2}[\varphi(t), \varphi(t)] = 0$
(2) $\varphi(0) = 0$

(3) $\left(\dfrac{\partial M_t}{\partial t}\right)_{t=0} \longrightarrow \eta = -\left(\dfrac{\partial\varphi(t)}{\partial t}\right)_{t=0} \in \Gamma(\bar\partial A^0(T)).$

Proof. (2) has already been done. As for (1), by Equation (2)

$$\left(\sum_{\beta} \varphi^{\beta}(t) \partial_{\beta} \zeta_j^{\alpha} - \bar{\partial} \zeta_j^{\alpha} \right) = 0.$$

So

$$0 = \bar{\partial} \left(\sum_{\beta} \varphi^{\beta}(t) \partial_{\beta} \zeta_j^{\alpha} \right)$$

$$= \sum_{\beta} \bar{\partial} \varphi^{\beta}(t) \partial_{\beta} \zeta_j^{\alpha} - \sum_{\beta} \varphi^{\beta}(t) \wedge \bar{\partial} \partial_{\beta} \zeta_j^{\alpha}.$$

Thus,

$$\sum_{\beta} \bar{\partial} \varphi^{\beta}(t) \partial_{\beta} \zeta_j^{\alpha} = \sum_{\beta} \varphi^{\beta}(t) \wedge \partial_{\beta} \bar{\partial} \zeta_j^{\alpha}$$

$$= \sum_{\beta} \varphi^{\beta}(t) \wedge \partial_{\beta} \left(\sum_{\gamma} \varphi(t)^{\gamma} \partial_{\gamma} \zeta_j^{\alpha} \right)$$

$$= \sum_{\beta} \sum_{\gamma} \varphi^{\beta}(t) \wedge \partial_{\beta} \varphi(t)^{\gamma} \cdot \partial_{\gamma} \zeta_j^{\alpha}$$

$$+ \sum_{\beta} \sum_{\gamma} \varphi^{\beta}(t) \wedge \varphi(t)^{\gamma} \partial_{\beta} \partial_{\gamma} \zeta_j^{\alpha}.$$

The last term in this expression is zero, since $\varphi^{\beta}(t) \wedge \varphi^{\gamma}(t)$ is skew-symmetric in β, γ and $\partial_{\beta} \partial_{\gamma} \zeta_j^{\alpha}$ is symmetric in β, γ. So we have

$$\sum_{\beta} \bar{\partial} \varphi^{\beta}(t) \partial_{\beta} \zeta_j^{\alpha} = \sum_{\gamma, \beta} \varphi(t)^{\gamma} \wedge \partial_{\gamma} \varphi^{\beta}(t) \partial_{\beta} \zeta_j^{\alpha}. \tag{6}$$

Now $\det(\partial_{\beta} \zeta_j^{\alpha}(z, t))$ is nonzero if t is small, so $\partial_{\beta} \zeta_j^{\alpha}$ is invertible for small t. Then (6) yields

$$\bar{\partial} \varphi^{\beta}(t) = \sum_{\gamma} \varphi^{\gamma}(t) \wedge \partial_{\gamma} \varphi^{\beta}(t)$$

$$= \tfrac{1}{2} [\varphi(t), \varphi(t)]^{\beta},$$

since

$$[\varphi, \varphi]^{\beta} = \sum \varphi^{\alpha} \wedge \partial_{\alpha} \varphi^{\beta} - (-1)^1 \varphi^{\alpha} \wedge \partial_{\alpha} \varphi^{\beta}$$

$$= 2 \sum \varphi^{\alpha} \wedge \partial_{\alpha} \varphi^{\beta}.$$

Finally, for (3), we already know

$$\eta = - \sum_{\beta} \bar{\partial} \dot{\zeta}_j^{\beta} \left(\frac{\partial}{\partial \zeta_j^{\beta}} \right)$$

where

$$\dot{\zeta}_j^{\beta} = \left(\frac{\partial \zeta_j^{\beta}}{\partial t} \right)_{t=0}.$$

But

$$\bar{\partial}\dot{\zeta}_j^{\alpha} = \sum_{\beta} \dot{\varphi}^{\beta}\partial_{\beta}\zeta_j^{\alpha} + \sum_{\beta} \varphi^{\beta}(0)\partial_{\beta}\dot{\zeta}_j^{\alpha};$$

and the last term is zero, so

$$\eta = -\sum_{\alpha,\beta} \dot{\varphi}^{\beta}\frac{\partial\zeta_j^{\alpha}}{\partial z^{\beta}}\left(\frac{\partial}{\partial\zeta_j^{\alpha}}\right)$$

$$= -\sum_{\beta}\dot{\varphi}^{\beta}\left(\frac{\partial}{\partial z^{\beta}}\right)$$

$$= -\left(\frac{\partial\varphi(t)}{\partial t}\right)_{t=0}. \qquad \text{Q.E.D.}$$

We wish to find conditions on a cohomology class ρ for it to represent an infinitesimal deformation. We first check to see that the bracket [,] extends to cohomology. Let

$$\varphi \in \Gamma(\bar{\partial}A^{0,\,p-1}(T)),$$

$$\psi \in \Gamma(\bar{\partial}A^{0,\,q-1}(T)),$$

so that

$$\bar{\partial}\varphi = \bar{\partial}\psi = 0.$$

Then it is obvious that $\bar{\partial}[\varphi, \psi] = 0$. Also if $\psi = \bar{\partial}\sigma$, then

$$[\varphi, \psi] = [\varphi, \bar{\partial}\sigma] = \pm\bar{\partial}[\varphi, \sigma].$$

Hence, [,] induces a map

$$H^p(M, \Theta) \otimes H^q(M, \Theta) \xrightarrow{\;[,]\;} H^{p+q}(M, \Theta)$$

by using Dolbeault, and the facts just noticed,

$$H^p(M, \Theta) \cong \frac{\Gamma(\bar{\partial}A^{0,\,p-1}(T))}{\bar{\partial}\Gamma(A^{0,\,p-1}(T))}.$$

Now let $(t_1, \cdots, t_m)$ be coordinates on B_r. Then any infinitesimal deformation is a linear combination of ones of the form

$$\eta_\nu = \left(\frac{\partial\varphi(t)}{\partial t_\nu}\right)_{t=0} = -\left(\frac{\partial M_t}{\partial t_\nu}\right)_{t=0}.$$

We claim:

THEOREM 1.2.   If $\rho \in H^1(M, \Theta)$ is an infinitesimal deformation, then

$$[\rho, \rho] = 0.$$

Proof.   By using the previous remark we need only check that

$$[\eta_\nu, \eta_\lambda] = 0$$

for all λ, ν. We differentiate the equation

$$\bar{\partial}\varphi(t) - [\varphi(t), \varphi(t)] = 0$$

twice at $t = 0$ to get

$$\bar{\partial}\left(\frac{\partial^2 \varphi(t)}{\partial t_\lambda\, \partial t_\nu}\right)_{t=0} = \left[\frac{\partial^2 \varphi}{\partial t_\lambda t_\nu}, \varphi(0)\right] + \left[\varphi(0), \frac{\partial^2 \varphi}{\partial t_\lambda t_\nu}\right]$$

$$+ 2\left[\frac{\partial \varphi}{\partial t_\lambda}, \frac{\partial \varphi}{\partial t_\nu}\right]$$

$$= 2\left[\frac{\partial \varphi}{\partial t_\lambda}, \frac{\partial \varphi}{\partial t_\nu}\right].$$

Let

$$\sigma_{\lambda\nu} = \frac{1}{2}\left(\frac{\partial^2 \varphi(t)}{\partial t_\lambda\, \partial t_\nu}\right)_{t=0}.$$

Then we get

$$[\eta_\lambda, \eta_\nu] = \bar{\partial}\,\sigma_{\lambda\nu} \sim 0. \qquad \text{Q.E.D.}$$

We remark that this is not a sufficient condition, and the higher derivatives give more information. It is quite difficult to compute $[\rho, \rho]$ for a general $\rho \in H^1(M, \Theta)$, but in most specific examples we get $[\rho, \rho] = 0$.

2.   An Existence Theorem for Deformations I. (No Obstructions)

We aim to prove the following theorem:

THEOREM 2.1.   [Kodaira, Nirenberg, and Spencer (1958)].   Let M be a compact complex manifold. Assume that $H^2(M, \Theta) = 0$. Then there exists a complex analytic family $\mathcal{M} \overset{\bar{\omega}}{\to} B_\varepsilon$, where

$$B_\varepsilon = \{t|\, |t| < \varepsilon\} \subseteq \mathbb{C}^m,\ m = \dim H^1(M, \Theta),$$

such that:

(1) $M_0 = \bar{\omega}^{-1}(0) = M$.

(2)   The map $T_0(B_\varepsilon) \to H^1(M, \Theta)$ given by $(\partial/\partial t) \to (\partial M_t/\partial t)_{t=0}$ is surjective (in fact, an isomorphism). More specifically if $\{\beta_1, \cdots, \beta_m\}$ is a base

for $H^1(M, \Theta)$, then we can define $\mathscr{M}$ so that

$$\left(\frac{\partial M_t}{\partial t_\nu}\right)_{t=0} = \beta_\nu \qquad \text{for } \nu = 1, \cdots, m.$$

Proof. We will accomplish the proof in the following two steps:

(1) Construction of a vector $(0, 1)$-form

$$\varphi(t) = \sum \varphi_{k_1 \cdots k_m} t_1^{k_1} \cdots t_m^{k_m}$$

such that

$$\varphi(0) = 0,$$
$$\bar{\partial}\varphi(t) - \tfrac{1}{2}[\varphi(t), \varphi(t)] = 0,$$

and

$$\left(\frac{\partial \varphi(t)}{\partial t_\nu}\right)_{t=0} = \beta_\nu \in H^1(M, \Theta).$$

(2) Show that $\varphi(t)$ determines a complex analytic family by using the Newlander-Nirenberg theorem.

First we survey the Newlander-Nirenberg theorem, which is sometimes called a " complex " Frobenius theorem. Let $U \subseteq \mathbb{C}^n$ be an open domain, and

$$\varphi = \sum \varphi_\alpha^\beta \, d\bar{z}^\alpha \left(\frac{\partial}{\partial z^\beta}\right)$$

a vector $(0, 1)$-form on U. Let

$$L_{\bar{\alpha}} = \left(\frac{\partial}{\partial \bar{z}^\alpha}\right) - \sum_{\beta=1}^{n} \varphi_\alpha^\beta(z)\left(\frac{\partial}{\partial z^\beta}\right).$$

We want to consider solutions to the equations

$$L_{\bar{\alpha}} f(z) = 0 \tag{1}$$

on the domain U. The theorem [of Newlander and Nirenberg (1957)] is:

THEOREM. If $L_{\bar{\alpha}}$ and $\bar{L}_{\bar{\alpha}}$ are (complex) linearly independent, and if

$$\bar{\partial}\varphi - \tfrac{1}{2}[\varphi, \varphi] = 0,$$

then Equation (1) has n C^∞ solutions $f_1(z), \cdots, f_n(z)$ such that

$$\det\left(\frac{\partial(f_1, \cdots, f_n, \bar{f}_1, \cdots, \bar{f}_n)}{\partial(z_1, \cdots, z_n, \bar{z}_1, \cdots, \bar{z}_n)}\right) \neq 0$$

(that is, $f_1, \cdots, f_n$ define a differentiable coordinate system on U).

REMARK 1.   If t is small, $\varphi(t)$ is small and $L_{\bar{\alpha}}$, $\bar{L}_{\bar{\alpha}}$ will be linearly independent.

REMARK 2.   Linear independence is needed; for if

$$L_{\bar{\alpha}} = \frac{\partial}{\partial x^{\alpha}} = \frac{\partial}{\partial \bar{z}^{\alpha}} + \frac{\partial}{\partial z^{\alpha}} \qquad (\varphi_{\bar{\alpha}}^{\beta} = -\delta_{\alpha}^{\beta}),$$

then $L_{\bar{\alpha}} f = 0$ implies f is independent of x^{α}.

REMARK 3.   If M is a complex manifold and φ is given satisfying the conditions of the theorem, then by using (the proof of) Proposition 1.2 we see M has another structure as a complex manifold which is described by the form φ. We say the almost complex structure φ is integrable, and hence associated to a complex structure.

In order to construct our form $\varphi(t)$ we need to do some more potential theory. We want to define the Green's operator on

$$\mathscr{L}^q = \Gamma(A^{0,q}(T)) = \text{the space of vector } (0, q)\text{-forms.}$$

To do this we introduce an Hermitian metric $g_{\alpha\beta}$ on M, and define an inner product,

$$(\varphi, \psi) = \int_M \sum g_{\alpha\beta}\, \varphi^{\alpha} \wedge * \psi^{\bar{\beta}},$$

where the $*$ operator has been defined before. We have the adjoint ϑ of $\bar{\partial}$, $(\vartheta\varphi, \psi) = (\varphi, \bar{\partial}\psi)$; and the Laplacian $\square = \vartheta\bar{\partial} + \bar{\partial}\vartheta$. Then the space of harmonic forms

$$\mathbb{H}^q = \{\varphi \mid \varphi \in \mathscr{L}^q,\ \square\varphi = 0\}$$
$$\cong H^q(M, \Theta),$$

defines a Hodge decomposition,

$$\mathscr{L}^q = \mathbb{H}^q \oplus \square\mathscr{L}^q = \mathbb{H}^q \oplus \bar{\partial}\mathscr{L}^{q-1} + \vartheta\mathscr{L}^{q+1}$$

into an orthogonal direct sum of subspaces. Thus, for $\varphi \in \mathscr{L}^q$, $\varphi = \eta + \square\psi$,

$\eta \in \mathbb{H}^q$, $\psi \in \mathscr{L}^q$. Since $\psi \in \mathscr{L}^q$,

$$\psi = \zeta + \psi_1,\ \zeta \in \mathbb{H}^q,\ \psi_1 \in \square\mathscr{L}^q,$$

and

$$\square\psi = \square\psi_1.$$

Thus

$$\varphi = \eta + \square\,\psi_1,\ \eta \in \mathbb{H}^q,\ \psi_1 \perp \mathbb{H}^q. \tag{2}$$

LEMMA 2.1. The decomposition in Equation (2) is unique.

Proof. Surely η is unique. If ψ', ψ are both orthogonal to $\mathbb{H}^q$ and

$$\varphi = \eta + \square\psi',$$
$$\varphi = \eta + \square\psi,$$

then

$$\square(\psi' - \psi) = 0$$

and

$$\psi' - \psi \in \mathbb{H}^q.$$

But

$$\psi' - \psi \perp \mathbb{H}^q \text{ so } \psi' = \psi. \qquad \text{Q.E.D.}$$

DEFINITION 2.1. Given φ, the unique ψ_1 making Equation (2) true is denoted $G\varphi$, and the mapping $\varphi \to G\varphi$ defines $G : \mathscr{L}^q \to \square\mathscr{L}^q$. G is called the *Green's operator*, and is a linear map. We write $\eta = H\varphi$ and call H the harmonic projection operator. Then

$$\varphi = H\varphi + \square G\varphi. \tag{3}$$

PROPOSITION 2.1. $\bar{\partial}H = H\bar{\partial} = 0$, $\vartheta H = H\vartheta = 0$, $GH = HG = 0$, $\bar{\partial}G = G\bar{\partial}$, $\vartheta G = G\vartheta$.

Proof $\bar{\partial}H\varphi = 0$ since $H\varphi \in \mathbb{H}^q = \{\psi \mid \bar{\partial}\psi = \vartheta\psi = 0\}$.

$H\bar{\partial}\varphi = 0$ since $\bar{\partial}\varphi \in \bar{\partial}\mathscr{L}^q \perp \mathbb{H}^q$. The proof that $0 = \vartheta H = H\vartheta$ is analogous. $HG\varphi = 0$ since $G\varphi \perp \mathbb{H}^q$. For $GH = 0$ notice

$$H\varphi = HH\varphi = H\varphi + \square GH_\varphi$$

and uniqueness yields $GH\varphi = 0$. The proofs of the last two are similar to each other so we only prove the first of them. Recall

$$\bar{\partial}\square = \bar{\partial}(\bar{\partial}\vartheta + \vartheta\bar{\partial}) = \bar{\partial}\vartheta\bar{\partial}$$

and

$$\square\bar{\partial} = \bar{\partial}\vartheta\bar{\partial}.$$

Thus $\bar{\partial}\square = \square\bar{\partial}$ and

$$\bar{\partial}\varphi = \bar{\partial}\square\, G\varphi = \square\bar{\partial}G\varphi$$
$$= H\bar{\partial}\varphi = \square G\bar{\partial}\varphi.$$

Since $\bar{\partial} G\varphi \perp \mathbb{H}^q$ and $H\bar{\partial}\varphi = 0$, we use uniqueness of decomposition Equation (3) to see $\bar{\partial} G\varphi = G\bar{\partial}\varphi$.   Q.E.D.

To proceed further, we need to introduce the Hölder norms in the spaces $\mathscr{L}^q$. To do this we fix a finite covering $\{U_j\}$ of M such that (z_j) are coordinates on U_j. Let $\varphi \in \mathscr{L}^q$,

$$\varphi = \sum_\lambda \varphi_j^\lambda(z)\left(\frac{\partial}{\partial z_j^\lambda}\right)$$

$$\varphi_j^\lambda = \frac{1}{q!} \sum \varphi_{j\bar{\alpha}_1 \cdots \bar{\alpha}_q}^\lambda \, d\bar{z}_j^{\alpha_1} \wedge \cdots \wedge d\bar{z}_j^{\alpha_q}.$$

Let $k \in \mathbb{Z}, k \geq 0; \alpha \in \mathbb{R}, 0 < \alpha < 1$. Let $h = (h_1, \cdots, h_{2n}), h_i \geq 0, \sum_{i=1}^{2n} h_i = |h|$ where $n = \dim M$. Then denote

$$D_j^h = \left(\frac{\partial}{\partial x_j^1}\right)^{h_1} \cdots \left(\frac{\partial}{\partial x_j^{2n}}\right)^{h_{2n}}, \qquad z_j^\alpha = x_j^{2\alpha-1} + ix_j^\alpha.$$

Then the Hölder norm $\|\varphi\|_{k+\alpha}$ is defined as follows:

$$\|\varphi\|_{k+\alpha} = \max_j\left\{ \sum_{\substack{h \\ |h| \leq k}} \left(\sup_{z \in U_j} |D_j^h \varphi_{j\bar{\alpha}_1 \cdots \bar{\alpha}_q}^\lambda(z)|\right) \right.$$
$$\left. + \sup_{\substack{y, z \in U_j \\ |h| = k}} \frac{|D_j^h \varphi_{j\bar{\alpha}_1 \cdots \bar{\alpha}_q}^\lambda(y) - D_j^h \varphi_{j\bar{\alpha}_1 \cdots \bar{\alpha}_q}^\lambda(z)|}{|y - z|^\alpha} \right\}, \tag{4}$$

where the sup is over all $\lambda, \alpha_1, \cdots, \alpha_q$. We have the following a priori estimate of Douglis and Nirenberg (1955).

$$\|\varphi\|_{k+\alpha} \leq C(\|\Box\varphi\|_{k-2+\alpha} + \|\varphi\|_0), \tag{5}$$

where $k \geq 2$, C is a constant which is independent of φ and

$$\|\varphi\|_0 = \max_{\substack{j \\ \lambda, \alpha_1, \cdots, \alpha_q}} \sup_{z \in U_j} |\varphi_{j\bar{\alpha}_1 \cdots \bar{\alpha}_q}^\lambda(z)|.$$

REMARK.   One can see that two norms defined as in Equation (4) for two different coverings $\{U_j\}$, $\{U_j'\}$ induce equivalent topologies on $\mathscr{L}^q$.

PROPOSITION 2.2. $\|[\varphi, \psi]\|_{k+\alpha} \leq C \|\varphi\|_{k+1+\alpha} \|\varphi\|_{k+1+\alpha}$, where C is independent of φ and ψ.

Proof.   We leave the simple check to the reader.

We need to know the following strong kind of continuity for the Green's operator G:

PROPOSITION 2.3. $\|G\varphi\|_{k+\alpha} \le C \|\varphi\|_{k-2+\alpha}$, $k \ge 2$, where C depends only on k and α, not on φ.

Proof. We use Equation (5) to get

$$\|G\varphi\|_{k+\alpha} \le C(\|\Box\varphi\|_{k-2+\alpha} + \|G\varphi\|_0)$$
$$\le C(\|\varphi - H\varphi\|_{k-2+\alpha} + \|G\varphi\|_0)$$
$$\le C(\|\varphi\|_{k-2+\alpha} + \|H\varphi\|_{k-2+\alpha} + \|G\varphi\|_0).$$

The space $\mathbb{H}^q$ is finite dimensional, so let $\{b_\nu\}_{\nu=1}^m$ be a base. Then

$$H\varphi = \sum_{\nu=1}^m (\varphi, \beta_\nu)\beta_\nu,$$

so

$$\|\varphi\|_{k-2+\alpha} \le \sum_{\nu=1}^m \|\beta_\nu\|_{k-2+\alpha} \max_\nu |(\varphi, \beta_\nu)|$$
$$\le C_1(\sum \|\beta_\nu\|_{k-2+\alpha})\|\varphi\|_0$$
$$\le C_2 \|\varphi\|_{k-2+\alpha}.$$

Thus,

$$\|G\varphi\|_{k+\alpha} \le C_3(\|\varphi\|_{k-2+\alpha} + \|G\varphi\|_0) \tag{6}$$

and we need only prove

$$\|G\varphi_0\| \le C_4 \|\varphi\|_{k-2+\alpha},$$

that is,

$$\frac{\|G\varphi\|_0}{\|\varphi\|_{k-2+\alpha}} \le C_4. \tag{7}$$

Suppose (7) is not true. Then there is a sequence $\varphi^{(\nu)}$ such that

$$\lim_\nu \frac{\|G\varphi^{(\nu)}\|_0}{\|\varphi^{(\nu)}\|_{k-2+\alpha}} = +\infty.$$

By multiplying $\varphi^{(\nu)}$ by a constant we may assume that $\|G\varphi^{(\nu)}\|_0 = 1$, and then $\|\varphi^{(\nu)}\|_{k-2+\alpha} \to 0$. Then Equation (6) implies that

$$\|G\varphi^{(\nu)}\|_{k+\alpha} \le K \qquad \text{(is bounded for } k \ge 2\text{)}. \tag{8}$$

Write

$$G\varphi^{(\nu)} = \frac{1}{g!} \sum (G\varphi^{(\nu)})^\lambda_{j\bar\alpha_1 \cdots \bar\alpha_q} \, dz_j^{\bar\alpha_1} \wedge \cdots \wedge dz_j^{\bar\alpha_q}.$$

Then Equation (8) implies that each of

$$(G\varphi^{(v)})^{\lambda}_{j\bar\alpha_1 \cdots \bar\alpha_q}$$

and all of its partial derivatives up to order k are uniformly bounded and equicontinuous. We are in a position to use Ascoli's theorem. We can choose a subsequence $\{\varphi^{(v_n)} \mid n = 1, 2, \cdots\}$ such that $G\varphi^{(v_n)}$ and $D^h_j G\varphi^{(v_n)}$ converge uniformly to ψ and $D^h_j \psi$ for $|h| \le k$. For simplicity let us replace v_n by v. Then we get

$$(\psi, \psi) = \lim_{v \to \infty}(G\varphi^{(v)}, \psi)$$

$$= \lim_{v \to \infty}(G\varphi^{(v)}, \Box G\psi + H\psi)$$

$$(\text{since } G\varphi^{(v)} \perp \mathbb{H}^q) = \lim_{v \to \infty}(G\varphi^{(v)}, \Box G\psi)$$

$$(\text{self adjointness}) = \lim_{v \to \infty}(\Box G\varphi^{(v)}, G\psi)$$

$$= \lim_{v \to \infty}(\varphi^{(v)} - H\varphi^{(v)}, G\psi)$$

$$(\text{since } \varphi^{(v)} \to 0) = 0.$$

Thus $\psi = 0$. But we should have $\|\psi\|_0 = 1$, since

$$\|\psi\|_0 = \lim_{v \to \infty} \|G\varphi^{(v)}\|_0 = 1.$$

This contradiction proves the proposition.

Let us now begin to construct the $\varphi(t)$ of Part (a). We use power series techniques but we notice that we could also use the implicit function theorem for Banach spaces [compare Kuranishi (1965)]. We want to construct $\varphi(t) = \sum_{\mu=1}^{\infty} \varphi_\mu(t)$, where

$$\varphi_\mu(t) = \sum_{v_1 + \cdots + v_m = \mu} \varphi_{v_1 \cdots v_m} t_1^{v_1} \cdots t_m^{v_m}$$

and each $\varphi_{v_1 \cdots v_m} \in \Gamma(A^{0,1}(T))$ such that

$$\bar\partial\varphi(t) - \tfrac{1}{2}[\varphi(t), \varphi(t)] = 0, \tag{9}$$

$$\varphi_1(t) = \sum_{v=1}^{m} \eta_v t_v, \tag{10}$$

where $\{\eta_v\}$ is a base for $\mathbb{H}^1 \cong H^1(M, \Theta)$. We use a method due to Kuranishi. Consider the equation

$$\varphi(t) = \varphi_1(t) + \tfrac{1}{2}\bar\partial G[\varphi(t), \varphi(t)], \tag{11}$$

where $\varphi_1(t)$ is given by (10). We first show that (11) has a unique formal power series solution $\varphi(t)$. In fact, this is clear since

$$\varphi_2(t) = \tfrac{1}{2}\vartheta G[\varphi_1(t), \varphi_1(t)]$$

$$\varphi_3(t) = \tfrac{1}{2}\vartheta G([\varphi_1(t), \varphi_2(t)] + [\varphi_2(t), \varphi_1(t)])$$

$$\varphi_\mu(t) = \tfrac{1}{2}\vartheta G\left(\sum_{\lambda=1}^{\mu-1} [\varphi_\lambda(t), \varphi_{\mu-\lambda}(t)]\right). \tag{12}$$

PROPOSITION 2.4.   For small $|t|$, $\varphi(t) = \sum_{\mu=1}^{\infty} \varphi_\mu(t)$ converges in the norm $\|\ \|_{k+\alpha}$.

Proof.   Let

$$A(t) = \frac{\beta}{16\gamma} \sum_{\mu=1}^{\infty} \gamma^\mu (t_1 + \cdots + t_m)^\mu$$

$$= \sum A_{v_1 \cdots v_m} t_1^{v_1} \cdots t_m^{v_m}.$$

As usual, $\|\varphi(t)\|_{k+\alpha} \ll A(t)$ means $\|\varphi_{v_1 \cdots v_m}\|_{k+\alpha} \le A_{v_1 \cdots v_m}$, and $\varphi^\mu(t) = \varphi_1(t) + \cdots + \varphi_\mu(t)$. Then (12) is equivalent to

$$\varphi^\mu(t) = \tfrac{1}{2}\vartheta G[\varphi^{\mu-1}(t), \varphi^{\mu-1}(t)] \bmod t^{\mu+1}. \tag{13}$$

We want to choose β and γ. Suppose they are chosen so that $\|\varphi^{\mu-1}(t)\|_{k+\alpha} \ll A(t)$. Since ϑ is a linear differential operator of first order,

$$\|\tfrac{1}{2}\vartheta G[\varphi, \psi]\|_{k+\alpha} \le C_1 \|G[\varphi, \psi]\|_{k+1+\alpha}$$

$$\le C_1 C_{k,\alpha} \|[\varphi, \psi]\|_{k-1+\alpha}$$

$$\le C_1 C_{k,\alpha} C \|\varphi\|_{k+\alpha} \|\psi\|_{k+\alpha},$$

by Propositions 2.2 and 2.3, where C_1, $C_{k,\alpha}$, and C are constants independent of φ and ψ. Hence by (13)

$$\|\varphi^{(\mu)}(t)\|_{k+\alpha} \le C_1 C_{k,\alpha} C \|\varphi^{(\mu-1)}(t)\|_{k+\alpha} \|\varphi^{(\mu-1)}(t)\|_{k+\alpha}$$

$$\ll C_1 C_{k,\alpha} C (A(t))^2$$

$$\ll C_1 C_{k,\alpha} C\left(\frac{\beta}{\gamma}\right) A(t)$$

as in Section 3, Chapter 2. Thus choose β and γ so that $C_1 C_{k,\alpha} C(\beta/\gamma) < 1$. Then $\|\varphi^\mu(t)\|_{k+\alpha} \ll A(t)$. These constants are all independent of μ. So if β and γ are chosen so that $\|\varphi^1(t)\|_{k+\alpha} \ll A(t)$, which is clearly possible, then $C_1 C_{k,\alpha} C(\beta/\gamma) < 1$ yields

$$\|\varphi(t)\|_{k+\alpha} \ll A(t).$$

So for small $|t|$, $\varphi(t)$ converges.   Q.E.D.

PROPOSITION 2.5. The $\varphi(t)$ of Proposition 2.4 satisfies $\bar{\partial}\varphi(t) - \frac{1}{2}[\varphi(t), \varphi(t)] = 0$ if and only if $H[\varphi(t), \varphi(t)] = 0$, where $H : \Gamma(A^{0,2}(T)) \to \mathbb{H}^2 \cong H^2(M, \Theta)$ is the orthogonal projection to the harmonic subspace of $\mathscr{L}^2 = \Gamma(A^{0,2}(T))$.

Proof. If $\bar{\partial}\varphi = \frac{1}{2}[\varphi, \varphi]$, then $0 = H\bar{\partial}\varphi = \frac{1}{2}H[\varphi, \varphi]$, since $H\bar{\partial} = 0$. Conversely let $H[\varphi, \varphi] = 0$ and set $\psi(t) = \bar{\partial}\varphi(t) - \frac{1}{2}[\varphi(t), \varphi(t)]$. Then each $\eta_\nu \in \mathbb{H}^1$ so $\bar{\partial}\eta_\nu = 0$ and

$$2\psi(t) = \bar{\partial}\vartheta G[\varphi(t), \varphi(t)] - [\varphi(t), \varphi(t)].$$

Recall that any ω can be decomposed $\omega = H\omega + \square G\omega$. Since $H[\varphi, \varphi] = 0$, we get $2\psi(t) = (\bar{\partial}\vartheta G - \square G)[\varphi(t), \varphi(t)]$. Because $\square = \bar{\partial}\vartheta + \vartheta\bar{\partial}$ we get

$$2\psi(t) = -\vartheta\bar{\partial}G[\varphi(t), \varphi(t)]$$
$$= -\vartheta G\bar{\partial}[\varphi(t), \varphi(t)]$$
$$= -2\vartheta G[\bar{\partial}\varphi(t), \varphi(t)].$$

This last equality is true because

$$\bar{\partial}[\varphi, \varphi] = [\bar{\partial}\varphi, \varphi] - [\varphi, \bar{\partial}\varphi]$$
$$= [\bar{\partial}\varphi, \varphi] + [\bar{\partial}\varphi, \varphi] = 2[\bar{\partial}\varphi, \varphi].$$

Then

$$\varphi(t) = -\vartheta G[\bar{\partial}\varphi(t), \varphi(t)]$$
$$= -\vartheta G[\psi(t) + \frac{1}{2}[\varphi(t), \varphi(t)], \varphi(t)]$$
$$= -\vartheta G[\psi(t), \varphi(t)]$$

by the Jacobi identity. Estimating, we get

$$\|\psi(t)\|_{k+\alpha} \le C_1 C_{k,\alpha} C \|\psi(t)\|_{k+\alpha} \|\varphi(t)\|_{k+\alpha}.$$

Choose $|t|$ so small that $\|\varphi(t)\|_{k+\alpha} C_1 C_{k,\alpha} C < 1$. Then we get the contradiction $\|\psi(t)\|_{k+\alpha} < \|\psi(t)\|_{k+\alpha}$ unless $\psi(t) = 0$ for all small t.   Q.E.D.

With the assumption $H^2(M, \Theta) = 0$ we have completed Part (a) of our task. We should notice the dependence of φ on z and t.

PROPOSITION 2.6. $\varphi(z, t)$ is C^∞ in (z, t) and holomorphic in t.

Proof. It is immediate that φ is C^k since the series converges in $\| \quad \|_{k+\alpha}$. C^∞ dependence is not so obvious. To give the proof we refer to the regularity theorem for quasi-linear elliptic operators [Douglis and Nirenberg (1955)].

Since the η_ν are harmonic,

$$\Box\varphi = \tfrac{1}{2}\Box\vartheta G[\varphi(t),\,\varphi(t)] = \tfrac{1}{2}\vartheta\Box G[\varphi(t),\,\varphi(t)]$$
$$= -\tfrac{1}{2}\vartheta H[\varphi(t),\,\varphi(t)] + \tfrac{1}{2}\vartheta[\varphi(t),\,\varphi(t)]$$
$$= \tfrac{1}{2}\vartheta[\varphi(t),\,\varphi(t)].$$

Since φ is holomorphic in t,

$$\sum_{\nu=1}^{m}\frac{\partial^2\varphi(t)}{\partial t_\nu\,\partial \bar t_\nu} = 0.$$

Thus φ satisfies

$$\sum_{\nu=1}^{m}\frac{\partial^2\varphi(t)}{\partial t_\nu\,\partial \bar t_\nu} + \Box\varphi(t) - \tfrac{1}{2}\vartheta[\varphi(t),\,\varphi(t)] = 0. \tag{14}$$

Equation (14) is quasi-linear elliptic since

$$\sum_{\nu=1}^{m}\frac{\partial^2}{\partial t_\nu\,\partial \bar t_\nu} + \Box$$

is elliptic and ϑ is first order. We know that $\varphi(z,\,t)$ is small for small t and that the coefficients of (14) are C^∞. Under these circumstances the regularity theorem says that solutions of (14) such as $\varphi(z,\,t)$ are C^∞.   Q.E.D.

So with the assumption that $H^2(M,\,\Theta) = 0$, the Newlander-Nirenberg theorem implies that each $\varphi(t)$ defines a complex structure M_t on M. Thus we obtain a family $\{M_t|\,|t| < \varepsilon\}$ of complex manifolds.

PROPOSITION 2.7. $\{M_t|\,|t| < \varepsilon\}$ is a complex analytic family.

Proof.   Consider $\varphi(t) = \sum \varphi_{\bar\alpha}^\beta(z,\,t)\,dz^{\bar\alpha}(\partial/\partial z^\beta)$ as a vector $(0,\,1)$-form defined on $M \times B_\varepsilon$ where $B_\varepsilon = \{t|\,|t| < \varepsilon\}$. Then $\varphi(t)$ satisfies the integrability condition $0 = \bar\partial\varphi(t) - \tfrac{1}{2}[\varphi(t),\,\varphi(t)]$ on $M \times B_\varepsilon$ where

$$\bar\partial\varphi(t) = \sum_{\nu=1}^{m}\frac{\partial\varphi(t)}{\partial\bar t^\nu}\,d\bar t^\nu + \sum_{\alpha=1}^{n}\frac{\partial\varphi(t)}{\partial\bar z^\alpha}\,d\bar z^\alpha$$

and $\partial\varphi(t)/\partial\bar t^\nu = 0$ since φ is holomorphic in t. Thus φ determines a complex structure $\mathcal{M}$ on $M \times B_\varepsilon$. The local complex coordinates of $\mathcal{M}$ are solutions ζ of

$$\bar\partial\zeta - \sum \varphi^\beta(t)\frac{\partial\zeta}{\partial z^\beta} = 0. \tag{15}$$

Equation (15) is satisfied if and only if

$$\frac{\partial}{\partial \bar{z}^\alpha} \zeta - \sum_\beta \varphi_\alpha^\beta \frac{\partial \zeta}{\partial z^\beta} = 0, \qquad \alpha = 1, \cdots, n$$

$$\frac{\partial \zeta}{\partial \bar{t}^\nu} = 0, \qquad \nu = 1, \cdots, m. \tag{16}$$

Hence, on some coordinate chart $\mathscr{U}_j = U_j \times B_\varepsilon \subseteq \mathscr{M}$ we have $n + m$ independent solutions $t_1, \cdots, t_m, \zeta_j^1(z, t), \cdots, \zeta_j^n(z, t)$ of Equation (15). So $\mathscr{M}$ is a complex manifold such that the projection $\pi : \mathscr{M} \to B_\varepsilon$ is holomorphic of rank m and for each fixed t, $\pi^{-1}(t) = M_t$ is the complex manifold with complex structure given by M_t.   Q.E.D.

This also completes the proof of Theorem 2.1. We remark that by the completeness theorem (see Chapter 2, Section 3) the family $\pi : \mathscr{M} \to B_\varepsilon$ is complete at $0 \in B_\varepsilon$.

3.   An Existence Theorem for Deformations II. (Kuranishi's Theorem)

We want to discuss the case in which $H^2(M, \Theta)$ does not necessarily vanish. Again fix an Hermitian metric on M and define ϑ, $\sqcap$, G, $\cdots$, and so forth. Let $\{\eta_\nu \mid \nu = 1, \cdots, m\}$ be a base for $\mathbb{H}^1 \cong H^1(M, \Theta)$. It is not necessary to know $H^2(M, \Theta) = 0$ for Proposition 2.4 to hold. Thus we still have a unique convergent (even in the norms defined later in this section) power series solution $\varphi(t)$ of

$$\varphi(t) = \eta(t) + \tfrac{1}{2}\vartheta G[\varphi(t), \varphi(t)], \tag{1}$$

where $\eta(t) = \sum_{\nu=1}^m \eta_\nu t_\nu$. And this $\varphi(t)$ satisfies

$$\bar{\partial}\varphi(t) - \tfrac{1}{2}[\varphi(t), \varphi(t)] = 0 \tag{2}$$

if and only if $H[\varphi(t), \varphi(t)] = 0$. Let $\{\beta_\lambda \mid \lambda = 1, \cdots, r\}$ be an orthonormal base of $\mathbb{H}^2$ and let $(,)$ be the inner product in $A^2 = \Gamma(A^{0,2}(T))$ (note the change of notation from Section 2, $A^2 \leftrightarrow \mathscr{L}^2$). Then

$$H[\varphi(t), \varphi(t)] = \sum_{\lambda=1}^r ([\varphi(t), \varphi(t)], \beta_\lambda)\beta_\lambda. \tag{3}$$

Hence $H([\varphi, \varphi]) = 0$ if and only if $([\varphi(t), \varphi(t)], \beta_\lambda) = 0$ for $\lambda = 1, \cdots, r$. Since $\varphi(t)$ is a power series in t so is $([\varphi(t), \varphi(t)], \beta_\lambda) = b_\lambda(t)$. Thus $b_\lambda(t)$ is holomorphic in t for $\lambda = 1, \cdots, r$ and $|t|$ small ($|t| < \varepsilon$). Also $b_\lambda(0) = 0$. Define an analytic set S as follows:

$$S = \{t \mid |t| < \varepsilon, b_\lambda(t) = 0, \lambda = 1, \cdots, r\}.$$

Then S is an analytic subset of B_ε containing the origin and we have proved the following proposition:

PROPOSITION 3.1. $\varphi(t)$ satisfies $\bar\partial\varphi(t) - \frac{1}{2}[\varphi(t), \varphi(t)] = 0$ if and only if $t \in S$.

COROLLARY. $\{M_t \mid t \in S\}$ is a set of complex structures on M.

We note that S may be singular; however, $\{M_t \mid t \in S\}$ can be made into a complex analytic family over S. The details are in Kuranishi (1965).

We want to prove that if ψ is a small enough vector $(0, 1)$-form such that $\bar\partial\psi - \frac{1}{2}[\psi, \psi] = 0$, then M_ψ is biholomorphically equivalent to one of the M_t, $t \in S$. This is the completeness theorem (an abbreviated version) of Kuranishi.

We find it convenient to use the Sobolev norms since some of the continuity properties that will be needed will be more transparent in these norms. For an open set U of $\mathbb{R}^n$ and complex C^∞ functions f and g defined on $\bar U$, the closure of U, we set

$$\langle f, g \rangle_k = \sum_{|\alpha| \le k} \int_U D^\alpha f(x) \cdot \overline{D^\alpha g(x)}\, dx, \tag{4}$$

where we use the multi-index notation $\alpha = (\alpha_1, \cdots, \alpha_n)$, $\alpha_i \ge 0$, $|\alpha| = \sum_{i=1}^n \alpha_i$, and $D^\alpha = (\partial/\partial x^1)^{\alpha_1} \cdots (\partial/\partial x^n)^{\alpha_n}$. Then

$$\|f\|_k = \sqrt{\langle f, f \rangle_k} = \|f\|_k^U. \tag{5}$$

The classical Sobelev lemma states that if V is a relatively compact open subset of U, then there is a constant c such that if $x \in V$ and $k > n/2$

$$|D^\alpha f(x)| \le c\, \|f\|_{k+|\alpha|}^U, \tag{6}$$

where c is a constant depending only on k, $|\alpha|$, V, U. As a consequence of (6) we have:

LEMMA 3.1.   There is a constant c such that if $k \ge n + 2$,

$$\|fg\|_k^V \le c\, \|f\|_k^U \cdot \|g\|_k^U. \tag{7}$$

Proof.   Let $|\ell| \le k$. Then
$$D^\ell(fg) = \sum_{r+s=\ell} D^r f \cdot D^s g.$$

Then either $|r| + n/2 + 1 \le k$ or $|s| + n/2 + 1 \le k$ when $r + s = \ell$, since $|\ell| \le k$ nd $k \ge n + 2$ implies $|\ell| + n + 2 \le 2k$. Thus, there is a constant κ so that

$$|D^\ell fg(x)| \le \kappa \sum_{r+s=\ell} \left(\|f\|_k^U |D^s g(x)| + |D^r f(x)| \|g\|_k^U \right)$$

for $x \in V$, by Equation (6). Squaring,

$$|D^\ell f g(x)|^2 \leq \kappa' \left[\sum_{r+s=\ell} (\|f\|_k^U)^2 |D^s g(x)|^2 + (\|g\|_k^U)^2 |D^r f(x)|^2 \right].$$

Now (7) follows easily.   Q.E.D.

By using a partition of unity, one defines $\|\varphi\|_k$ for any $\varphi \in A^P = \Gamma(A^{0,P}(T))$. The estimates of the previous section are essentially the same and we list the ones we need.

$$\|[\varphi, \psi]\|_k \leq c_k \|\varphi\|_{k+1} \|\psi\|_{k+1} \quad [\text{for all large } k. \ (k \geq 2n + 2 \text{ where } \dim_{\mathbb{C}} M = n).] \tag{8}$$

$$\|H\varphi\|_k \leq c_k \|\varphi\|_k \tag{9}$$

$$\|\vartheta g \varphi\|_k \leq c_k \|\varphi\|_{k-1}. \tag{10}$$

From now on k will be chosen so that the conditions of Lemma 3 hold.

PROPOSITION 3.2.   For fixed $\eta(t)$, Equation (1) has only one small ($\|\varphi\|_k < \varepsilon$) solution.

Proof.   Let $\tau = \varphi - \varphi(t)$. Then

$$\tau = \tfrac{1}{2}\vartheta G([\varphi, \varphi] - [\varphi(t), \varphi(t)])$$
$$= \tfrac{1}{2}\vartheta G([\tau, \varphi(t)] + [\varphi(t), \tau] + [\tau, \tau])$$
$$= \tfrac{1}{2}\vartheta G(2[\tau, \varphi(t)] + [\tau, \tau]).$$

Estimating $\|\tau\|_k$ gives

$$\|\tau\|_k \leq D(\|\tau\|_k \|\varphi(t)\|_k + \|\tau\|_k^2)$$
$$\leq D\|\tau\|_k(\|\varphi(t)\|_k + \|\tau\|_k).$$

If $\|\varphi(t)\|_k$ is small enough, the only way

$$0 \leq x \leq Dx(\|\varphi(t)\|_k + x)$$

can happen with x close to 0 is for $x = 0$.   Q.E.D.

The set $N = \{\eta(t) = \sum \eta_v t_v \mid |t| < \varepsilon\}$ describes a small neighborhood of $0 \in \mathbb{H}^1$. For small enough ε there is a one-to-one correspondence between $\eta \in N$ and solutions φ of (1). This is proved as follows: If $\varphi = \eta(t) + \tfrac{1}{2}\vartheta G[\varphi, \varphi]$, then $H\vartheta = 0$ implies $\eta = H\varphi$. Thus η is uniquely determined by φ. Given η, call the small solution of Equation (1) (given in Proposition 3.2) φ. Then a correspondence F is defined by $F\eta = \varphi$.

LEMMA 3.2. Suppose M_ψ is given, If $\vartheta\psi = 0$, then $\psi = \varphi(t)$ for some t if $\|\psi\|_k < \delta$ for small δ.

Proof.

$$\bar{\partial}\psi - \tfrac{1}{2}[\psi, \psi] = 0,$$

hence

$$\square\psi = \vartheta\bar{\partial}\psi + \bar{\partial}\vartheta\psi$$
$$= \tfrac{1}{2}\vartheta[\psi, \psi]$$

since $\vartheta\psi = 0$. Then

$$\psi - H\psi = G\square\psi = \tfrac{1}{2}G\vartheta[\psi, \psi].$$

Let $\eta = H\psi$. Then $\psi = \eta + \tfrac{1}{2}\vartheta G[\psi, \psi]$; and by assumption $\|\psi\|_k$ is small so that $\|\eta\|_k$ is small by Equation (9). Hence $\eta = \eta(t)$ for some $|t| < \varepsilon$, and the remark just made shows that $\psi = F\eta = F\eta(t) = \varphi(t)$.　　Q.E.D.

In general $\vartheta\psi \neq 0$ so we must try something else. When we say that M_ψ is holomorphically equivalent to $M_{\varphi(t)}$ we mean that there is a biholomorphic map $f: M_{\varphi(t)} \to M_\psi$ and $f: M \to M$ thus induces a diffeomorphism $f: M \to M$ of the underlying differentiable manifold M. Conversely let f be any diffeomorphism such that its values and first derivatives are close to the values and first derivatives of the identity map. Cover M_ψ with a system $\{U_j, \zeta_j^\alpha(z)\}$ of local ψ-holomorphic coordinates and let $\{U_j'\}$, $U_j' \subseteq U_j$ be a covering of M so that $f(U_j') \subseteq U_j$. Then $\zeta_j^\alpha(f(z))$ is a local holomorphic coordinate on U_j'. Thus φ is determined by the equation

$$\bar{\partial}\zeta^\alpha(f(z)) = \sum_{\beta=1}^n \varphi^\beta(z)\partial_\beta \zeta^\alpha(f(z)). \tag{11}$$

We also know

$$\bar{\partial}\zeta^\alpha(z) = \sum_{\beta=1}^n \psi^\beta(z)\partial_\beta \zeta^\alpha(z),$$

that is,

$$\frac{\partial}{\partial\bar{z}^\lambda}\zeta^\alpha(z) = \sum_{\beta=1}^n \psi_{\bar\lambda}^\beta(z)\frac{\partial\zeta^\alpha(z)}{\partial\bar{z}^\beta}. \tag{12}$$

Putting Equations (11) and (12) together we get

$$\sum_{\beta} \frac{\partial \zeta^{\alpha}(f(z))}{\partial f^{\beta}} \bar{\partial} f^{\beta}(z) + \sum_{\lambda} \frac{\partial \zeta^{\alpha}(f(z))}{\partial \bar{f}^{\lambda}} \bar{\partial} \bar{f}^{\lambda}$$

$$= \sum_{\gamma, \beta} \varphi^{\gamma}(z) \frac{\partial \zeta^{\alpha}}{\partial f^{\beta}} \partial_{\gamma} f^{\beta} + \sum_{\gamma, \lambda} \varphi^{\gamma}(z) \frac{\partial \zeta^{\alpha}}{\partial \bar{f}^{\lambda}} \partial_{\gamma} \bar{f}^{\lambda},$$

$$\sum_{\beta} \frac{\partial \zeta^{\alpha}}{\partial f^{\beta}} \left(\bar{\partial} f^{\beta} + \sum_{\lambda} \psi_{\lambda}^{\beta} \bar{\partial} \bar{f}^{\lambda} \right)$$

$$= \sum_{\beta, \gamma} \frac{\partial \zeta^{\alpha}}{\partial f^{\beta}} (\partial_{\gamma} f^{\beta} + \sum \psi_{\lambda}^{\beta} \partial_{\gamma} \bar{f}^{\lambda}) \varphi^{\gamma}(z).$$

The matrix $(\partial \zeta^{\alpha}/\partial f^{\beta})$ is invertible since it is assumed that $\|\psi\|_k$ is close to zero. Thus,

$$\bar{\partial} f^{\beta}(z) + \sum_{\lambda} \psi_{\lambda}^{\beta}(f(z)) \bar{\partial} \bar{f}^{\lambda}(z)$$

$$= \sum_{\gamma} \left(\partial_{\gamma} f^{\beta}(z) + \sum_{\lambda} \psi_{\lambda}^{\beta}(f(z)) \partial_{\gamma} \bar{f}^{\lambda}(z) \right) \varphi^{\gamma}(z). \tag{13}$$

PROPOSITION 3.3.   Let M_{φ} be the complex structure induced from M_{ψ} by the map $f: M \to M$. Then φ is determined by Equation (13). We use the notation $\psi \circ f$ to denote φ.

Proof.   We just notice that $\|\psi\|_k$ small implies that

$$\left(\partial_{\gamma} f^{\beta}(z) + \sum_{\lambda} \psi_{\lambda}^{\beta}(f(z)) \partial_{\gamma} \bar{f}^{\lambda}(z) \right)$$

is an invertible matrix.   Q.E.D.

Thus to prove Kuranishi's theorem in the case $\vartheta\psi \neq 0$ it suffices to show the existence of a diffeomorphism f such that $\vartheta(\psi \circ f) = 0$. This is our task. We need to digress for a moment to describe a way of indexing diffeomorphisms close to the identity by the use of geodesics.

We shall be brief, and refer the reader to any text on differential geometry for missing details. Let an Hermitian metric $(g_{\alpha\beta})$ be fixed on M. Then we have the Christoffel symbols

$$\Gamma_{\lambda\beta}^{\alpha} = \sum_{\mu} g^{\bar{\mu}\alpha} \left(\frac{\partial g_{\beta\bar{\mu}}}{\partial z^{\lambda}} \right),$$

and if $u(z) = \sum u^\alpha(z)(\partial/\partial z^\alpha)$ is a vector field, we have the covariant derivatives

$$\nabla_\lambda u^\alpha = \frac{\partial u^\alpha}{\partial z^\lambda} + \sum_\beta \Gamma^\alpha_{\lambda\beta} u^\beta,$$

$$\bar\nabla_\lambda u^\alpha = \frac{\partial u^\alpha}{\partial \bar z^\lambda}.$$

If $z(t)$ is a curve in M, we define

$$\nabla_t u^\alpha(z(t)) = \sum_\lambda \nabla_\lambda u^\alpha \frac{dz^\lambda(t)}{dt} + \sum \bar\nabla_\lambda u^\alpha \frac{\overline{dz^\lambda(t)}}{dt}$$

$$= \frac{d}{dt} u^\alpha(z(t)) + \sum_{\lambda,\beta} \Gamma^\alpha_{\lambda\beta}(z(t)) \frac{dz^\lambda(t)}{dt} u^\beta(z(t)),$$

when $u(z)$ is defined along $z(t)$. Then the geodesics of M are the curves $z(t)$ satisfying the differential equation

$$\nabla_t\left(\frac{dz^\alpha(t)}{dt}\right) = 0,$$

that is,

$$\frac{d(z^\alpha(t))}{dt^2} + \sum_{\lambda,\beta} \Gamma^\alpha_{\lambda\beta}(z(t)) \frac{dz^\lambda}{dt}(t) \frac{dz^\beta}{dt}(t) = 0. \tag{14}$$

Since M is compact, all metrics are *complete*, that is, the geodesics through a given point with given tangent direction are defined for all time t. Let $z^\alpha(t) = z^\alpha(t, z_0, \xi)$ be the solution of Equation (14) satisfying $z^\alpha(0) = z_0^\alpha$, $(dz^\alpha/dt)(0) = \xi^\alpha$. Then the following facts are easily verified by using existence and uniqueness theorems of ordinary differential equations:

(1) $z^\alpha(t, z_0, \xi)$ are C^∞ in (t, z_0, ξ).
(2) $z^\alpha(kt, z_0, \xi) = z^\alpha(t, z_0, k\xi)$.

We set $f^\alpha(z_0, \xi) = z^\alpha(1, z_0, \xi)$. Then f is C^∞ in (z_0, ξ) and $f^\alpha(z_0, t\xi) = z^\alpha(t, z_0, \xi)$. Differentiating this relation we get

$$\frac{dz^\alpha}{dt}(t, z_0, \xi) = \sum_{\beta=1}^n \xi^\beta \frac{\partial f^\alpha}{\partial \xi^\beta}(z_0, t\xi) + \sum_{\beta=1}^n \bar\xi^\beta \frac{\partial f^\alpha}{\partial \bar\xi^\beta}(z_0, t\xi).$$

So

$$\xi^\alpha = \frac{dz^\alpha}{dt}(0, z_0, \xi) = \sum_{\beta=1}^n \left[\xi^\beta \frac{\partial f^\alpha}{\partial \xi^\beta}(z_0, 0) + \bar\xi^\beta \frac{\partial f^\alpha}{\partial \bar\xi^\beta}(z_0, 0)\right].$$

This implies

$$f^\alpha(z_0, 0) = z_0^\alpha, \quad \frac{\partial f^\alpha}{\partial \xi^\beta}(z_0, 0) = \delta^\alpha_\beta, \quad \frac{\partial f^\alpha}{\partial \bar\xi^\beta}(z_0, 0) = 0.$$

The Taylor expanison then yields

$$f^{\alpha}(z_0, \xi) = z_0^{\alpha} + \xi^{\alpha} + 0(|\xi|^2), \tag{15}$$

where $0(|\xi|^2)$ is a term bounded by $M|\xi|^2$ for some $M > 0$ and for small $|\xi|$. Given a vector field $\xi = \sum_{\alpha=1}^{n} \xi^{\alpha}(z)(\partial/\partial z^{\alpha})$ on M we define a diffeomorphism $f_{\xi}: M \to M$ by

$$z^{\alpha} \to f^{\alpha}(z, \xi(z)).$$

By Equation (15)

$$f_{\xi}^{\alpha}(z) = z^{\alpha} + \xi^{\alpha}(z) + 0(|\xi(z)|^2). \tag{16}$$

We wish to calculate $\varphi = \psi \circ f_{\xi}$ and we use Equation (13). We abbreviate (16) with $f_{\xi}^{\alpha} = z^{\alpha} + \xi^{\alpha} + r^{\alpha}$. Then (13) becomes

$$\bar{\partial}\xi^{\beta} + \bar{\partial}r + \sum_{\lambda} \psi_{\lambda}^{\beta}(f_{\xi})(d\bar{z}^{\lambda} + \bar{\partial}\xi^{\lambda} + \bar{\partial}\bar{r}^{\lambda})$$

$$= \sum_{\gamma=1}^{n} \left[\delta_{\lambda}^{\beta} + \partial_{\gamma}\xi^{\beta} + \partial_{\gamma}r^{\beta} + \sum_{\lambda} \psi_{\lambda}^{\beta}(f_{\xi})(\partial_{\gamma}\bar{\xi}^{\lambda} + \partial_{\gamma}\bar{r}^{\lambda}) \right]\varphi^{\gamma}.$$

Multiplying by the inverse of the expression in brackets $[-]$ we get

$$\varphi^{\gamma} = \bar{\partial}\xi^{\gamma} + \sum_{\lambda} \psi_{\lambda}^{\gamma}(f_{\xi}) \, d\bar{z}^{\lambda} + \cdots$$

$$= \bar{\partial}\xi^{\gamma} + \sum_{\lambda} \psi_{\lambda}^{\gamma}(z) \, d\bar{z}^{\lambda} + R^{\gamma}(\psi, \xi).$$

Thus,

$$\psi \circ f_{\xi} = \varphi = \bar{\partial}\xi + \psi + R(\psi, \xi), \tag{17}$$

where $R(t\psi, t\xi) = t^2 R_1(\psi, \xi, t)$ if t is a real number and both R, R_1 are C^{∞} functions of the parameters $\psi_{\beta}^{\alpha}(z)$, $\psi_{\beta}^{\alpha}(f_{\xi}(z))$, $\xi^{\alpha}(z)$, $(\partial\xi^{\alpha}/\partial z^{\beta})(z)$, $(\partial\xi^{\alpha}/\partial\bar{z}^{\beta})(z)$ in local coordinates.

In A^0 we have H^0, the space of holomorphic vector fields on M. Using the L_2 inner product on A^0 we let F^0 be the orthogonal complement of H^0. So $\xi \in F^0$ if and only if $(\xi, \eta) = 0$ for all $\eta \in H^0$, that is, F^0 is the kernel of the map $H: A^0 \to H^0$. Then for $\xi \in F^0$,

$$\xi = G\square\xi + H\xi = G\square\xi.$$

Since ϑ is zero on A^0, $\vartheta\xi = 0$, and

$$\square\xi = \vartheta\bar{\partial}\xi,$$

yielding

$$\xi = G\vartheta\bar{\partial}\xi. \tag{18}$$

Now give A^0, A^1 and their subspaces the $\| \; \|_k$ topology.

PROPOSITION 3.4. There are neighborhoods of the origin U and V in A^1 and F^0, respectively, so that for any $\psi \in U$ there is a unique $\xi = \xi(\psi)$ in V such that

$$\vartheta(\psi \circ f_\xi) = 0. \tag{19}$$

Proof. Using (17) we see (19) is satisfied if and only if

$$0 = \vartheta(\psi \circ f_\xi) = \vartheta\bar{\partial}\xi + \vartheta\psi + \vartheta R(\psi, \xi).$$

By (18)

$$\xi = G\vartheta\bar{\partial}\xi = -G\vartheta\psi - G\vartheta R(\psi, \xi).$$

Thus (19) is equivalent to

$$\xi + G\vartheta\psi + G\vartheta R(\psi, \xi) = 0. \tag{20}$$

Now choosing neighborhoods U_1, V_1 so that R is defined on $U_1 \times V_1$ we can define a map

$$h: U_1 \times V_1 \to F^0$$

by $h(\psi, \xi) = \xi + G\vartheta\psi + G\vartheta R(\psi, \xi)$. By our previous remarks on $R(\psi, \xi)$, we see that h is continuous if U_1, V_1, F_0 have the $\| \quad \|_k$ topology, since R is continuous as a map from $U_1 \times V_1$ with the $\| \quad \|_k$ topology to A^1 with the $\| \quad \|_{k-1}$ topology. In fact, h is even uniformly continuous and hence has a (unique) extension to a mapping h of the completion of the domain to the completion of F^0. Then the partial derivative

$$\left. \frac{\partial h}{\partial \xi} \right|_{(0,\,0)} : \hat{F}^0 \longrightarrow \hat{F}^0,$$

where $\hat{F}^0$ is the completion of F^0, is the identity map. Then by the implicit function theorem [see Lang (1962)] there is a C^∞ function g on a small neighborhood of the origin in the space $\hat{A}^1$ (completion of A^1) such that Equation (20) is satisfied if and only if $\xi = g(\psi)$ for some $\psi \in U$. Thus given $\psi \in U$, where U is a small neighborhood in A^1, there is a unique solution $g(\psi) = \xi$ of (20) which is sufficiently small. Then $\square + \vartheta R(\psi, \quad) + \vartheta$ is an elliptic second-order equation with C^∞. Thus, if $\psi \in U \subseteq A^1$, $\xi = g(\psi)$ satisfies

$$\square\xi + \vartheta R(\psi, \xi) + \vartheta\psi = 0$$

so ξ is C^∞, that is, $\xi \in F^0$.   Q.E.D.

Let us summarize our conclusions.

THEOREM 3.1. (Kuranishi)

(a) Let M be a given compact complex manifold, and let $\{\eta_v\}$ be a base for $\mathbb{H}^1 \cong H^1(M, \Theta)$. Let $\varphi(t)$ be the solution of the equation

$$\varphi(t) = \eta(t) + \tfrac{1}{2}\vartheta G[\varphi(t), \varphi(t)],$$

where $\eta(t) = \sum_{v=1}^{m} t_v \eta_v$, $|t| < \rho$, and let

$$B = \{t \mid H[\varphi(t), \varphi(t)] = 0\}.$$

Then for each $t \in B$, $\varphi(t)$ determines a complex structure M_t on M.

(b)   Let ψ be any vector $(0, 1)$-form satisfying $\bar{\partial}\psi - \frac{1}{2}[\psi, \psi] = 0$. Then ψ defines a complex structure M_ψ on M. If $\|\psi\|_k$ is small enough, there is a unique $\xi \in F^0$ such that $\psi \circ f_\xi = \varphi(t)$ for some $t \in B$, and hence M_ψ is biholomorphically equivalent to M_t.

4.   Stability Theorems

The main point of this section is to prove that if $\mathcal{M} = \{M_t\}$ is a complex analytic family and M_{t_0} is Kähler, then M_t is Kähler if $|t - t_0|$ is small. We first study elliptic differential equations depending on a parameter.

Let $P = \{t \mid |t| < \alpha, t = (t_1, \cdots, t_r)\}$ be an open disk in $\mathbb{C}^r$. Let X be a compact differentiable manifold, and let $\mathcal{B}$ be a differentiable complex vector bundle over $X \times P$. Let $B_t = \mathcal{B}|_{X \times \{t\}}$ be the restriction of $\mathcal{B}$ to $X \times \{t\}$. Then we can consider $\mathcal{B} = \{B_t\}$ to be a family of vector bundles over X depending differentiably on $t \in P$. Let

$$L(\mathcal{B}) = \text{the space of differentiable sections of } \mathcal{B},$$
$$L(B_t) = \text{the space of differentiable sections of } B_t.$$

We are only interested in small deformations so we can assume that P is small and then there will be a finite covering $\{X_i\}$ of X such that

$$\mathcal{B}|_{X_i \times P} = \mathbb{C}^\mu \times X_i \times P,$$

that is, $\mathcal{B}$ is trivial over $X_i \times P$. Let (ζ_i^λ, x, t) be a local coordinate on $\mathcal{B}|_{X_i \times P}$. Then the coordinate transformations on $\mathcal{B}$ are written as

$$\zeta_i^\lambda = \sum_{v=1}^{\mu} b_{ikv}^\lambda(x, t)\zeta_k^v.$$

By an (even-order) *differential operator*

$$E_t : L(B_t) \longrightarrow L(B_t),$$

we mean a map which can be written locally in the form

$$(E_t \psi)_i^\lambda(x) = \sum_{v=1}^{\mu} E_{iv}^\lambda(x, t, D_i)\psi_i^v(x),$$

where $\psi(x) = (\psi_i^\lambda(x))$ is a section of B_t and where $E_{iv}^\lambda(x, t, D_i)$ is a polynomial of degree $m \equiv 0 \pmod 2$ in $D_i = (\partial/\partial x_i^\alpha)$. In our applications we will only need $m = 2, 4$. For E_t to be well defined we must have

$$\sum_{\tau, v} E_{iv}^\lambda(x, t, D_i)b_{ik\tau}^v(x, t)\psi_k^\tau(x) = \sum_{\tau, v} b_{ikv}^\lambda(x, t)E_{k\tau}^v(x, t, D_k)\psi_k^\tau(x),$$

where

$$\psi_i^\lambda(x) = \sum_{\nu=1}^{\mu} b_{ik\nu}^\lambda(x, t)\psi_k^\nu(x).$$

We say that E_t depends differentiably on t if all of the coefficients of $E_{i\nu}^\lambda(x, t, D_i)$ as a polynomial in D_i are C^∞ functions of (x, t). We assume given a Riemannian metric on M and an Hermitian metric on the fibres of B_t so that

$$\sum_{\lambda, \nu} g_{i\lambda\bar\nu}(x,t)\, \zeta_i^\lambda\zeta_i^{\bar\nu}\, dX_i = \sum_{\sigma, \tau} g_{k\sigma\bar\tau}(x, t)\zeta_k^\sigma\zeta_k^{\bar\tau}\, dX_k$$

is invariantly defined. We have written these expressions in terms of local coordinates for X where

$$\zeta_i^\lambda = \sum_\sigma b_{ik\sigma}^\lambda(x, t)\zeta_k^\sigma$$

and

$$dX_i = dx_i^1 \cdots dx_i^n \qquad (n = \dim X).$$

We assume that $g_{i\lambda\bar\nu}(x, t)$ are C^∞ in (x, t). An inner product on $L(B_t)$ is defined by

$$(\varphi, \psi)_t = \int_X \sum g_{i\lambda\bar\nu}(x, t)\varphi_i^\lambda(x)\overline{\psi_i^\nu}(x)\, dX_i. \tag{1}$$

We will consider only those E_t which are formally self-adjoint; that is,

$$(E_t\varphi, \psi)_t = (\varphi, E_t\psi)_t \tag{2}$$

for all $\varphi, \psi \in L(B_t)$. Let us see what this implies for the coefficients of E_t. Let $E_{i\nu}^{m\lambda}(x, t, D_i)$ be the terms of order m in $E_{i\nu}^\lambda(x, t, D_i)$. Writing out (2) we get

$$\int_X \sum_{\lambda, \tau, \nu} g_{i\lambda\tau} E_{i\nu}^{m\lambda}(x, t, D_i)\varphi_i^\nu(x)\overline{\psi_i^\tau}(x) + \text{lower order terms}$$

$$= \int_X \sum_{\lambda, \nu, \tau} g_{i\lambda\bar\nu} \varphi_i^\lambda(x)\overline{E_{i\tau}^{m\nu}(x, t, D_i)\psi_i^\tau}(x) + \text{lower order terms}.$$

Integrating the first term of the left-hand side by parts and using the fact that m is even we get

$$\int_X g_{i\lambda\bar\nu} \varphi_i^\lambda\overline{E_{i\tau}^{m\nu}}\psi^\tau = \int \sum g_{i\lambda\tau} \varphi_i^\nu E_{i\nu}^{m\lambda}\overline{\psi_i^\tau} + \text{lower order terms}.$$

This is true for all φ and ψ so we get

$$\sum_{\nu, \tau} g_{i\lambda\bar\nu} \overline{E_{i\tau}^{m\nu}}\psi_i^\tau = \sum g_{i\nu\bar\tau} E_{i\lambda}^{m\nu}\overline{\psi_i^\tau}$$

for all ψ. Hence,

$$\sum_\nu g_{i\nu\bar\tau} E_{i\lambda}^{m\nu} = \sum_\nu g_{i\lambda\bar\nu} \overline{E_{i\tau}^{m\nu}}. \tag{3}$$

Replace $(\partial/\partial x_i^\alpha)$ in $E_{i\lambda}^{m\nu}$ with y_α. Then $E_{i\lambda}^{m\nu}(x, t, y)$ is a homogeneous polynomial of order m in y with coefficients which are C^∞ functions of (x, t). Equation (3) becomes

$$\overline{A_{i\lambda\bar\tau}} = A_{i\tau\bar\lambda}$$

if $A_{i\lambda\bar\tau} = \sum_\nu g_{iv\bar\tau} E_{i\lambda}^{m\nu}(x, t, y)$. We have proved:

PROPOSITION 4.1. Equation (2) implies (3).

We shall assume further that E_t is strongly elliptic, that is,

$$(-1)^{m/2} \sum_{\lambda, \tau} A_{i\lambda\bar\tau}(x, t, y) w_\lambda \bar w_\tau > 0$$

for any *real* $y = (y_1, \cdots, y_n) \neq 0$, and any complex $w = (w_1, \cdots, w_\mu) \neq 0$. We need to collect some facts about such operators. We quote the following well-known theorem which can be found, for example, in Palais (1965, p. 182):

THEOREM 4.1. E_t has a complete orthonormal set of eignenfunctions $\{e_{th}\}_{h=1}^\infty \subseteq L(B_t)$. Let the eigenvalues be $\lambda_h(t)$. Then they are real and

$$E_t(e_{th}) = \lambda_h(t) e_{th}, \qquad \text{with } (e_{th}, e_{tj})_t = \delta_{ik}.$$

[Completeness means that any $\psi \in L(B_t)$ can be written as follows:

$$\psi = \sum_{h=1}^\infty a_j e_{th}, \qquad \text{where } a_h = (\psi, e_{th})_t.]$$

Furthermore we can arrange the e_{th} such that

$$\lambda_1(t) \leq \lambda_2(t) \leq \cdots$$

and $\lim_{h\to\infty} \lambda_h(t) = +\infty$.

The following theorem is proved in Kodaira and Spencer (1960), and we shall not prove it here.

THEOREM 4.2. Each eigenvalue $\lambda_h(t)$ is a continuous function of $t \in P$.

REMARK. $\lambda_h(t)$ may not be differentiable. For example, let

$$E_t = \begin{pmatrix} \alpha(t)\beta(t) \\ \gamma(t)\delta(t) \end{pmatrix}$$

be a 0th-order differential operator (just a matrix). Then

$$\lambda(t) = \frac{\alpha(t) + \delta(t) \pm \sqrt{(\alpha(t) - \delta(t))^2 + 4\beta(t)\,\delta(t)}}{2}$$

which could fail to be differentiable when $(\alpha - \delta)^2 + 4\beta\delta = 0$. Let $\mathbb{F}_t$ be the kernel of E_t. Then

$$\mathbb{F}_t = \{\psi \mid E_t \psi = 0\} = \left\{ \psi \mid \psi = \sum_{\lambda_h(t)=0} a_h \, e_{th} \right\}.$$

Let F_t be the orthogonal projection to $\mathbb{F}_t$, that is,

$$F_t \psi = \sum_{\lambda_h(t)=0} (\psi, e_{th}) e_{th} \,.$$

The Green's operator G_t is defined by

$$G_t \psi = \sum_{\lambda_h(t)\neq 0} \frac{1}{\lambda_h(t)} (\psi, e_{th}) e_{th} \,.$$

F_t and G_t are related by the equation

$$\psi = E_t G_t \psi + F_t \psi.$$

We have already investigated the case $P = $ a point, $E_t = \square$, $F_t = H$, $\mathbb{F}_t = \mathbb{H}$, $G_t = G$. In the general case we have the following theorem:

THEOREM 4.3. dim $\mathbb{F}_t$ is an upper semicontinuous function of t. This means that given t_0, there is a small enough ε so that dim $\mathbb{F}_t \leq$ dim $\mathbb{F}_{t_0}$ for $|t - t_0| < \varepsilon$.

Proof. dim $\mathbb{F}_t = d_t$ is finite since $\lambda_n(t) \to \infty$. In the ordering of the $\lambda_h(t_0)$ we have

$$\cdots \leq \lambda_j(t_0) < 0 = \lambda_{j+1}(t_0) = \cdots = \lambda_{j+d_0}(t_0) < \lambda_{j+d_0+1}(t) \leq \cdots .$$

By continuity, choose ε so that $\lambda_j(t) < 0$ and $\lambda_{j+d_0+1}(t) > 0$ if $|t - t_0| < \varepsilon$. Then in this disk $d_t \leq d_0$.   Q.E.D.

Next we want to say what we can about the differentiability of F_t and G_t.

DEFINITION 4.1. Given $\psi_t \in L(B_t)$ for each $t \in P$, we say that ψ_t depends differentiably on t if there is a $\psi \in L(\mathscr{B})$ such that

$$\psi_t = \psi \mid_{X \times \{t\}} \,.$$

Let $(\zeta^\alpha_{j,x,t})$ be a local coordinate on $\mathscr{B}$, where

$$\zeta^\lambda_j = \sum b^\lambda_{jk\nu}(x, t) \zeta^\nu_k.$$

Then if $\psi_t = \{\psi^\lambda_{tj}(x)\}$ in local coordinates, we have

$$\psi^\lambda_{tj}(x) = \sum b^\lambda_{jk\nu}(x, t) \psi^\nu_{tk}(x).$$

If $\psi \in L(\mathscr{B})$ is such that $\psi_t = \psi \mid_{X \times \{t\}}$, then $\psi = \{\psi^\lambda_j(x, t)\}$ and $\psi^\lambda_j(x, t) = \psi^\lambda_{tj}(x)$ so $\psi^\lambda_{tj}(x)$ are C^∞ in x and t. Conversely, if the $\psi^\lambda_{tj}(x)$ are C^∞ in x and t, it is easy to see that ψ_t depends differentiably on t.

DEFINITION 4.2.   Given a linear operator $A_t : L(B_t) \to L(B_t)$ for each $t \in P$, we say that A_t depends differentiably on t if the following condition is satisfied: If ψ_t depends differentiably on t, then $A_t \psi_t$ depends differentiably on t.

REMARK. E_t depends differentiably on t in this sense.

Given E_t, let $I = [\alpha, \beta]$ be a real interval on the λ-line such that $0 \in I$. We define

$$F_t(I)\psi = \sum_{\lambda_h(t) \in I} (\psi, e_{th})e_{th}$$

$$G_t(I)\psi = \sum_{\lambda_h(t) \notin I} (\psi, e_{th}) \frac{e_{th}}{\lambda_h(t)}.$$

THEOREM 4.4.   Let $t_0 \in P$. If $\lambda_h(t_0) \neq \alpha, \beta$ for all h, then there is an open set V around t_0 such that $F_t(I)$ and $G_t(I)$ depend differentiably on $t \in V$.

Sketch of Proof.   Consider a rectangular contour C in the ζ-plane as in the figure below.

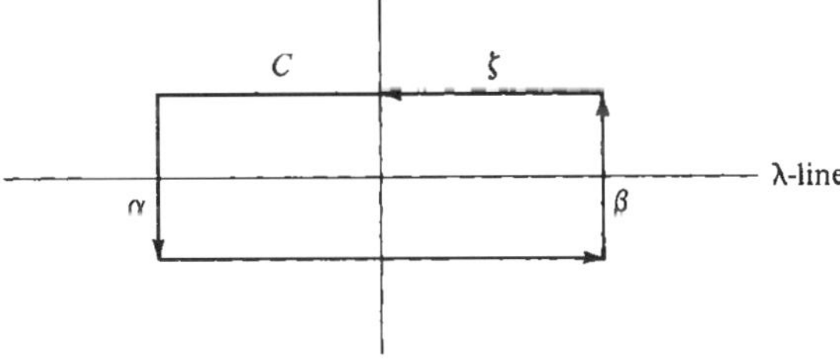

Figure 10

Let $\zeta \in C$ and form $E_t - \zeta$. Then we have

$$\|(E_t - \zeta)\psi\|_t^2 = \|\sum (\psi, e_{th})(\lambda_h(t) - \zeta)e_{th}\|_t^2$$
$$\geq \min_h |\lambda_h(t) - \zeta|^2 \|\psi\|_t^2$$
$$\geq k \|\psi\|_t^2,$$

where k is some constant (greater than zero), as long as t remains in a small compact neighborhood of t_0. This implies easily that $(E_t - \zeta)^{-1} = G_t(\zeta)$ is defined and continuous. It is harder to see that $G_t(\zeta)$ depends differentiably on (t, ζ). This is proved in Kodaira and Spencer (1960). We can write down $G_t(\zeta)$:

$$G_t(\zeta)\psi = \sum_{k=1}^{\infty} \frac{(\psi, e_{th})}{(\lambda_h(t) - \zeta)} e_{th}.$$

Then integrating around the contour C we get

$$F_t(I) = -\frac{1}{2\pi i} \int_C G_t(\zeta)\, d\zeta$$

because

$$\frac{1}{2\pi i} \int_C \frac{d\zeta}{\lambda - \zeta} = \begin{cases} -1, & \text{if } \lambda \in I \\ 0, & \text{if } \lambda \notin I \end{cases} \qquad (\lambda \text{ is real}).$$

We also have

$$G_t(I) = \frac{1}{2\pi i} \int_C G_t(\zeta)\, \frac{d\zeta}{\zeta}$$

since

$$\frac{1}{2\pi i} \int_C \frac{d\zeta}{(\lambda - \zeta)\zeta} = \begin{cases} 0, & \text{if } \lambda \in I \\ 1, & \text{if } \lambda \notin I. \end{cases}$$

Thus $G_t(I)$ and $F_t(I)$ depend differentiably on t.   Q.E.D.

THEOREM 4.5.   If dim $\mathbb{F}_t$ is independent of $t \in V$, where V is some open subset of P, then F_t and G_t depend differentiably on t.

Proof.   Since the $\lambda_h(t)$ are continuous in t and dim $\mathbb{F}_t$ is independent of t, we can find a small open set U around t and small interval $I = [-\varepsilon, +\varepsilon]$ around 0 such that $F_s = F_s(I)$ and $G_s = G_s(I)$ for $s \in U$. Then apply Theorem 4.4.

Before we apply these theorems, we need to make some more definitions.

DEFINITION 4.3.   By a *differentiable* family of compact complex manifolds we mean a triple $(\mathcal{M}, \bar\omega, P)$ where $\mathcal{M}, P$ are differentiable manifolds, $\bar\omega$ is a differentiable map of rank $=$ dim P, and each $M_t = \bar\omega^{-1}(t)$ is a complex manifold. More precisely for each $x_0 \in \mathcal{M}$ there should be a local diffeomorphism from $\mathcal{U}_j \ni x_0$

$$x \longrightarrow (z^1, \cdots, z^n, t^1, \cdots, t^m)$$

to a domain in $\mathbb{C}^n \times \mathbb{R}^m$, where $\bar\omega(x) = (t^1, \cdots, t^m)$ are coordinates around $\bar\omega(x)$, and such that for a fixed t, $(z^1(x), \cdots, z^n(x))$ is a complex coordinate on M_t. On $\mathcal{U}_j \cap \mathcal{U}_k$,

$$z^a_j = f^a_{jk}(z_k, t),$$

where f^a_{jk} is differentiable in (z_k, t) and holomorphic in z_k for a fixed t. Let $T(\mathcal{M})$ be the complex vector bundle over $\mathcal{M}$ whose transition functions on

$\mathcal{U}_j \cap \mathcal{U}_k$ are given by the matrix

$$\left(\frac{\partial z_j^\alpha}{\partial z_k^\beta} \right).$$

$T(\mathcal{M})$ is the *bundle of holomorphic tangent vectors along the fibres of $\mathcal{M}$*. Let $T^*(\mathcal{M})$ be its dual and $\bar{T}(\mathcal{M})$ its conjugate bundle. Then we set

$$\mathcal{T}^*(r, s) = \left(\underset{r}{\wedge} T^*(\mathcal{M}) \right) \wedge \left(\underset{s}{\wedge} \bar{T}^*(\mathcal{M}) \right).$$

We call $\mathcal{T}^*(r, s)$ the bundle of (r, s)-forms along the fibres. Let r_t be restriction to M_t. Then we have

$$r_t(T(\mathcal{M})) = T(M_t)$$

and

$$r_t(\mathcal{T}^*(r, s)) = T^{*(r, s)}(M_t).$$

Let $L^{r, s}$ be the space of C^∞ sections of $\mathcal{T}^*(r, s)$. Then $L_t^{r, s} = r_t L^{r, s}$ is the space of C^∞ (r, s)-forms on M_t. Any $\psi \in L^{r, s}$ can be written

$$\psi = \frac{1}{r! s!} \sum \psi_{\alpha_1 \cdots \alpha_r \, \bar{\beta}_1 \cdots \bar{\beta}_s}(z, t) \, dz_j^{\alpha_1} \wedge \cdots \wedge dz_j^{\alpha_r} \wedge d\bar{z}_j^{\beta_1} \wedge \cdots \wedge d\bar{z}^{\beta_2},$$

where $dz_j^\alpha = (\partial/\partial z_j^\alpha)^*$. Our transformation law for $(\partial/\partial z_j^\alpha)$ is

$$\left(\frac{\partial}{\partial z_k^\alpha} \right) = \sum_{\beta=1}^n \frac{\partial z_j^\beta}{\partial z_k^\alpha} \left(\frac{\partial}{\partial z_j^\beta} \right).$$

Thus the law for dz_j^α is

$$dz_j^\alpha = \sum_{\beta=1}^n \frac{\partial z_j^\alpha}{\partial z_k^\alpha} \, dz_k^\beta.$$

This implies that dz_j^α is *not* the differential of z_j^α. If it were, it would transform according to the law

$$dz_j^\alpha = \sum_{\beta=1}^n \frac{\partial z_j^\alpha}{\partial z_k^\beta} \, dz_k^\beta + \sum_{\lambda=1}^m \frac{\partial z_j^\alpha}{\partial t^\lambda} \, dt^\lambda.$$

Next, let $\mathcal{B}$ be a vector bundle on $\mathcal{M}$ and let $B_t = r_t(\mathcal{B})$. Then $L^{r, s}(\mathcal{B})$ is the space of C^∞ sections of $\mathcal{B} \otimes \mathcal{T}^*(r, s)$ and $L^{r, s}(B_t)$ is the space of C^∞ sections of $B_t \otimes T_t^{*(r, s)}(M_t)$. If $\mathcal{B}$ is given by the transition equation

$$\zeta_j^\lambda = \sum_{v=1}^\mu b_{jkv}^\lambda(z, t)\zeta_k^v$$

and $\psi \in L^{r, s}(\mathcal{B})$ is given locally by $\psi = (\psi_j^1, \cdots, \psi_j^\mu)$, then

$$\psi_j^\lambda = \sum_{v=1}^\mu b_{jkv}^\lambda(z, t)\psi_k^v,$$

where

$$\psi_j^\lambda = \frac{1}{r!\,s!} \sum \psi_{j\alpha_1 \cdots \bar\beta_1 \cdots \bar\beta_s}^\lambda(z, t)\, dz_j^{\alpha_1} \wedge \cdots \wedge d\bar z^{\beta_s}.$$

We have assumed an Hermitian metric on $\mathscr{B}$ depending differentiably on t; hence, if φ_t, ψ_t are in $L(B_t)$ and are C^∞ in t, then $(\varphi_t, \psi_t)_t$ depends differentiably on t. We can introduce the operators d_t, ∂_t, $\bar\partial_t$, ϑ_t, δ_t, $\bar\vartheta_t$ acting on $L_t^{r,\,s} = L^{r,\,s}(B_t)$. Then the operators

$$\Box_t = \bar\partial_t \vartheta_t + \vartheta_t \bar\partial_t, \quad \Delta_t = d_t \delta_t + \delta_t d_t, \quad \square_t$$

depend differentiably on t. We now state the main theorem of this section.

THEOREM 4.6.   Let $\mathscr{M} \xrightarrow{\;\overline\omega\;} P$ be a differentiable family. If $M_{t_0} = \overline\omega^{-1}(t_0)$ is a Kähler manifold, then $M_t = \overline\omega^{-1}(t)$ is Kähler for $|t - t_0|$ small enough.

REMARK 1.   A good problem would be to find an elementary proof (for example, using power series methods). Our proof uses nontrivial results from partial differential equations (Theorem 4.1).

REMARK 2.   Hironaka (1962) has given an example of a non-Kähler deformation of a Kähler manifold. Hence the theorem is only true for $|t - t_0|$ small.

Proof.   (of the theorem)   Assume $P = \{t \mid |t| < 1\} \subseteq \mathbb{R}^m$. Since M_0 is Kähler we have a Kähler form ω_0 on M_0

$$\omega_0 = i \sum g_{\alpha\bar\beta}(z)\, dz^\alpha \wedge dz^\beta.$$

We extend this to an Hermitian metric on all of the fibres as follows: Let $(z_j^1, \cdots, z_j^n, t^1, \cdots, t^m)$ be local coordinates on $\mathscr{U}_j \cong U_j \times P \subseteq \mathbb{C}^n \times \mathbb{R}^m$, such that $\overline\omega(z_j^1, \cdots, t^m) = (t^1, \cdots, t^m)$. Then on $\mathscr{U}_j$

$$i \sum_{\alpha,\,\beta} g_{j\alpha\bar\beta}(z_j)\, dz_j^\alpha \wedge d\bar z_j^\beta = \omega_j$$

is a C^∞ Hermitian form which is independent of t. Let $\{\rho_j(z, t)\}$ be a partition of unity subordinate to $\{\mathscr{U}_j\}$. Then we define

$$\omega_t = \sum_j \rho_j(z, t)\omega_j(z) = i \sum_{\alpha,\,\beta} g_{t\alpha\bar\beta}(z)\, dz^\alpha \wedge d\bar z^\beta.$$

The metric ω_t depends differentiably on t and ω_0 is the ω_0 we started with. Let our inner products $(\varphi, \psi)_t$ be defined with respect to the metric ω_t where $\varphi, \psi \in L_t^{r,\,s}$. Then the operator $\Box_t = \vartheta_t \bar\partial_t + \bar\partial_t \vartheta_t$ is strongly elliptic and depends differentiably on t so our theorems apply. In fact, one can check that the principal part of $\Box_t$ is

$$-\sum g_t^{\bar\beta\alpha} \frac{\partial^2}{\partial z^\alpha\, \partial\bar z^\beta}$$

in a local coordinate where $(g_t^{\bar\beta\alpha})$ is the inverse matrix to $(g_{t\alpha\bar\beta})$ (see, for example, Chapter 3, Section 2). We define the following spaces:

$$Z_{\bar\partial_t}^{r,s} = \{\varphi \mid \varphi \in L_t^{r,s}\,\bar\partial_t\,\varphi = 0\},$$

$$Z_{\partial_t}^{r,s} = \{\varphi \mid \varphi \in L_t^{r,s}, \partial_t\,\varphi = 0\},$$

$$L_t^q = \sum_{r+s=q} L_t^{r,s},$$

$$Z_{d_t}^q = \{\varphi \mid \varphi \in L_t^q, d_t\,\varphi = 0\},$$

$$\mathbb{H}_t^{r,s} = \{\varphi \mid \varphi = L_t^{r,s}, \square_t\,\varphi = 0\},$$

$$\mathbb{H}_t^q = \{\varphi \mid \varphi \in L_t^q, \Delta_t\,\varphi = 0\}.$$

The theorems of Dolbeault and de Rham yield

$$\mathbb{H}_t^{r,s} \cong \frac{Z_{\bar\partial_t}^{r,s}}{\bar\partial_t L_t^{r,s-1}} \cong H^s(M_t, \Omega_t^r), \tag{4}$$

$$\mathbb{H}_t^q \cong H^q(M_t, \mathbb{C}),$$

where Ω_t^r is the sheaf of germs of holomorphic r-forms on M_t. As usual, let

$$h_t^{r,s} = \dim \mathbb{H}_t^{r,s}, \ b^q = \dim \mathbb{H}_t^q.$$

Then b^q is the qth Betti number of M_t (which is independent of t since all of the M_t are diffeomorphic to each other). M_0 is Kähler, so

$$\Delta_0 = 2\square_0 = 2\bar\square_0. \tag{5}$$

This was proved in Chapter 3, Section 5. We have for each t

$$\bar\partial_t\partial_t + \partial_t\bar\partial_t = 0 = \bar\vartheta_t\partial_t + \vartheta_t\bar\vartheta_t, \tag{6}$$

since Kähler is not necessary for this. However, if the reader will consult the proof of (5) he will find that in the Kähler case (for example, $t = 0$)

$$\vartheta_0\partial_0 + \partial_0\vartheta_0 = 0 = \bar\vartheta_0\bar\partial_0 + \bar\partial_0\bar\vartheta_0. \tag{7}$$

Now we define

$$E_t = \partial_t\bar\partial_t\vartheta_t\bar\vartheta_t + \vartheta_t\bar\vartheta_t\partial_t\bar\partial_t + \vartheta_t\partial_t\bar\vartheta_t\bar\partial_t + \bar\vartheta_t\bar\partial_t\vartheta_t\partial_t + \vartheta_t\bar\partial_t + \bar\vartheta_t\partial_t.$$

Then

$$(E_t\varphi, \psi)_t = (\vartheta_t\bar\partial_t\varphi, \vartheta_t\bar\vartheta_t\psi)_t + (\partial_t\bar\partial_t\varphi, \partial_t\bar\partial_t\psi)_t$$

$$+ (\vartheta_t\partial_t\varphi, \vartheta_t\partial_t\psi)_t + (\bar\partial_t\varphi, \bar\partial_t\psi)_t + (\partial_t\varphi, \partial_t\psi)_t.$$

PROPOSITION 4.2. E_t *is a strongly elliptic self-adjoint differential operator of order 4 acting on* $L_t^{r,s}$.

Proof. We clearly have

$$(E_t\varphi, \psi)_t = (\varphi, E_t\psi)_t. \tag{8}$$

Thus E_t is self-adjoint. We let the reader check that the principal part of E_t is

$$\sum_{\alpha, \beta, \gamma, \delta} g_i^{\bar\beta\alpha} g_i^{\bar\delta\gamma} \frac{\partial^4}{\partial z_i^\alpha \, \partial \bar z_i^\beta \, \partial z_i^\gamma \, \partial \bar z_i^\delta}$$

in a local coordinate system.   Q.E.D.

PROPOSITION 4.3. $E_0 = \Box_0 \Box_0 + \vartheta_0 \bar\partial_0 + \bar\vartheta_0 \partial_0$ and $E_t \varphi = 0$ if and only if $\vartheta_t \bar\vartheta_t \varphi = \bar\partial_t \varphi = \partial_t \varphi = 0$.

Proof.   Using Equations (5), (6), and (7) we get

$$\begin{aligned}
\Box_0 \Box_0 &= (\bar\vartheta_0 \partial_0 + \partial_0 \bar\vartheta_0)(\vartheta_0 \bar\partial_0 + \bar\partial_0 \vartheta_0) \\
&= \bar\vartheta_0 \partial_0 \vartheta_0 \bar\partial_0 + \partial_0 \bar\vartheta_0 \bar\partial_0 \vartheta_0 + \partial_0 \bar\vartheta_0 \vartheta_0 \bar\partial_0 + \bar\vartheta_0 \partial_0 \bar\partial_0 \vartheta_0 \\
&= \vartheta_0 \bar\vartheta_0 \partial_0 \bar\partial_0 + \partial_0 \bar\partial_0 \vartheta_0 \bar\vartheta_0 + \vartheta_0 \partial_0 \bar\vartheta_0 \bar\partial_0 + \bar\vartheta_0 \bar\partial_0 \vartheta_0 \partial_0 .
\end{aligned}$$

This proves the first statement. For the second statement we use the obvious inequality

$$(E_t \varphi, \varphi) \geq \| \vartheta_t \bar\vartheta_t \varphi \|_t^2 + \| \bar\partial_t \varphi \|_t^2 + \| \partial_t \varphi \|_t^2 .   \text{Q.E.D.}$$

As before, let $\mathbb{F}_t^{r,s} = \{\varphi \mid E_t \varphi = 0, \varphi \in L_t^{r,s}\}$. Let $F_t : L_t^{r,s} \to \mathbb{F}_t^{r,s}$ be orthogonal projection to $\mathbb{F}_t^{r,s}$ and G_t be the Green's operator, so

$$\psi = E_t G_t \psi + F_t \psi.$$

PROPOSITION 4.4. $Z_{d_t}^{r,s} = \partial_t \bar\partial_t L_t^{r-1,s-1} \oplus \mathbb{F}_t^{r,s}$, where $\oplus$ means orthogonal direct sum.

Proof.   If $\varphi \in \mathbb{F}_t^{r,s}$, then $d_t \varphi = 0$. If $\psi \in \partial_t \bar\partial_t L_t^{r-1,s-1}$ then, $d_t \varphi = 0$ since $d_t \partial_t \bar\partial_t = (\partial_t + \bar\partial_t)\partial_t \bar\partial_t = 0$. Thus $\partial_t \bar\partial_t L_t^{r-1,s-1} + \mathbb{F}_t^{r,s} \subseteq Z_{d_t}^{r,s}$. The sum is orthogonal since

$$(\partial_t \bar\partial_t \varphi, \eta)_t = (\varphi, \vartheta_t \bar\vartheta_t \eta)_t = 0$$

for $\eta \in \mathbb{F}_t^{r,s}$.

Next take $\psi \in Z_t^{r,s}$. Then

$$\psi = E_t G_t \psi + F_t \psi = \partial_t \bar\partial_t \alpha + \vartheta_t \beta + \bar\vartheta_t \gamma + \eta,$$

where $\eta = F_t \psi$. Since $d_t \psi = 0$

$$d_t(\vartheta_t \beta + \bar\vartheta_t \gamma) = 0. \tag{9}$$

Let $\sigma = \vartheta_t \beta + \bar\vartheta_t \gamma$. Then $\sigma \in L_t^{r,s}$ and $\partial_t : L_t^{r,s} \to L_t^{r+1,s}$, $\bar\partial_t : L_t^{r,s} \to L_t^{r,s+1}$, so $\partial_t \sigma = \bar\partial_t \sigma = 0$. We claim $\sigma = 0$. For

$$(\sigma, \sigma)_t = (\vartheta_t \beta + \bar\vartheta_t \gamma, \sigma)_t = (\beta, \bar\partial_t \sigma)_t + (\gamma, \partial_t \sigma)_t = 0.$$

Thus $\sigma = 0$. Hence $\psi = \partial_t \bar{\partial}_t \alpha + \eta$, and

$$Z_t^{r,s} \subseteq \partial_t \bar{\partial}_t L_t^{r-1,s-1} \oplus \mathbb{F}_t^{r,s}. \qquad \text{Q.E.D.}$$

LEMMA 4.1. $\dim \mathbb{F}_t^{1,1} \geq b^2 - 2h_t^{0,2}$ where b^2 is the second Betti number of M_t.

Proof.

$$\mathbb{F}_t^{1,1} \cong Z_{d_t}^{1,1}/\partial_t \bar{\partial}_t L_t^0.$$

Since

$$\partial_t \bar{\partial}_t L_t^0 \subseteq dL_t^1 \cap Z_{d_t}^{1,1}$$

$$\dim \mathbb{F}_t^{1,1} \geq \dim\left(\frac{Z_{d_t}^{1,1}}{dL_t^1} \cap Z_{d_t}^{1,1}\right) = \dim\left(\frac{Z_{d_t}^{1,1} + dL_t^1}{d_t L_t^1}\right).$$

By de Rham's theorem $b^2 = \dim(Z_{d_t}^2/d_t L_t^1)$. We claim there is the following exact sequence

$$0 \longrightarrow \frac{Z_{d_t}^{1,1} + d_t L_t^1}{d_t L_t^1} \longrightarrow \frac{Z_{d_t}^2}{d_t L_t^1} \overset{\pi_t}{\longrightarrow} \frac{Z_{\partial_t}^{2,0}}{\partial_t L_t^{1,0}} + \frac{Z_{\bar{\partial}_t}^{0,2}}{\bar{\partial}_t L_t^{0,1}}.$$

We must define π_t and check that it has the correct kernel. Let $\psi \in Z_{d_t}^2$, $d_t \psi = 0$ where $\psi = \psi^{2,0} + \psi^{1,1} + \psi^{0,2}$. Then $d_t \psi = 0$ yields $\bar{\partial}_t \psi^{0,2} = \partial_t \psi^{2,0} = 0$. So we can map ψ to $\psi^{2,0} + \psi^{0,2} \in Z_{\partial_t}^{2,0} + Z_{\bar{\partial}_t}^{0,2}$. Let $\psi \in d_t L_t^1$. Then

$$\psi = d_t(\varphi^{1,0} + \varphi^{0,1}) \longrightarrow \partial_t \varphi^{1,0} + \bar{\partial}_t \varphi^{1,0}.$$

This correspondence induces the map π_t. To compute $\ker \pi_t$, suppose $\pi_t \psi = \psi^{2,0} + \psi^{0,2} = \partial_t \sigma^{1,0} + \bar{\partial}_t \tau^{0,1}$. Then

$$\psi - d_t(\sigma^{1,0} + \tau^{0,1}) = \xi = \psi^{1,1} - \bar{\partial}_t \sigma^{1,0} - \partial_t \tau^{0,1} = \xi^{1,1}$$

where ξ is of type $(1,1)$ Then

$$d_t \xi = d_t \psi = 0.$$

Thus $\xi \in Z_{d_t}^{1,1}$ and this yields $\psi \in d_t L_t^1 + Z_{d_t}^{1,1}$. This exact sequence implies

$$b^2 - \dim\left(\frac{Z_{d_t}^{1,1} + d_t L_t^1}{d_t L_t^1}\right) \leq \dim\left(\frac{Z_{\partial_t}^{2,0}}{\partial_t L_t^{1,0}}\right) + \dim\left(\frac{Z_{\bar{\partial}_t}^{0,2}}{\bar{\partial}_t L_t^{0,1}}\right).$$

But Dolbeault's theorem implies

$$\dim\left(\frac{Z_{\bar{\partial}_t}^{0,2}}{\bar{\partial}_t L_t^{0,1}}\right) = \dim H^2(M, \mathcal{O}) = h_t^{0,2}$$

and

$$h_t^{0,2} = h_t^{2,0} = \dim\left(\frac{Z_{\partial_t}^{2,0}}{\partial_t L_t^{1,0}}\right).$$

Thus $\dim \mathbb{F}_t^{1,1} \geq b^2 - 2h_t^{0,2}.$ Q.E.D.

LEMMA 4.2. $\dim \mathbb{F}_t^{1,1} = \dim \mathbb{F}_0^{1,0}$ for small $|t|$.

Proof.   Let us compute $\dim \mathbb{F}_0^{1,1}$. We claim

$$\mathbb{F}_0^{1,1} = H_0^{1,1} \cong H^1(M_0, \Omega^1). \tag{10}$$

We use (5) and Proposition 4.3 for the proof of (10). Since

$$(E_0\,\varphi,\,\varphi) = \|\square_0\,\varphi\|^2 + \|\partial_0\,\varphi\|^2 + \|\bar{\partial}_0\,\varphi\|^2,$$

$E_0\,\varphi = 0$ implies $\square_0\,\varphi = 0$. Conversely if $\square_0\,\varphi = 0$, then $\bar{\partial}_0\,\varphi = 0 = \vartheta_0\,\varphi$; and $\square_0\,\varphi = 0$ so $\partial_0\,\varphi = \bar{\vartheta}_0\,\varphi = 0$. Thus $E_0\,\varphi = 0$. This proves (10). Hence,

$$\dim \mathbb{F}_0^{1,1} = h_0^{1,1}.$$

Recall that on a Kähler manifold

$$b^2 = h_0^{2,0} + h_0^{1,1} + h_0^{0,2}$$
$$= h_0^{1,1} + 2h_0^{0,2}.$$

Thus,

$$\dim \mathbb{F}_0^{1,1} = b^2 - 2h_0^{0,2}.$$

By the upper semicontinuity of $\dim \mathbb{F}_t^{1,1}$,

$$\dim \mathbb{F}_t^{1,1} \le \dim \mathbb{F}_0^{1,1}, \qquad \text{for } |t| < \varepsilon.$$

Also by upper semicontinuity of $\dim \mathbb{H}_t^{0,2}$,

$$h_t^{0,2} \le h_0^{0,2}, \qquad \text{for } |t| < \varepsilon.$$

Thus,

$$\dim \mathbb{F}_0^{1,1} \ge \dim \mathbb{F}_t^{1,1} \ge b^2 - 2h_t^{0,2} \ge b^2 - h_0^{0,2} = \dim \mathbb{F}_0^{1,1}. \qquad \text{Q.E.D.}$$

COROLLARY. F_t depends differentiably on t.

Proof.   Use Proposition 4.5.

Now we shall finish the proof of Theorem 4.6. We have

$$\omega_t = i \sum_{\alpha,\,\beta} g_{\alpha\bar{\beta}}(z,\,t)\, dz^\alpha \wedge d\bar{z}^\beta,$$

which depends differentiably on t and ω_0 is a Kähler form M_0. Hence, $d\omega_0 = 0$. In fact, if the reader will consult Chapter 3, Section 5 he will find we proved that $\omega_0 \in \mathbb{H}_0^{1,1} = \mathbb{F}_0^{1,1}$. Thus $F_0 \omega_0 = \omega_0$. Since $\psi_t = F_t \omega_t$ depends differentiably on t, $\psi_t \to \omega_0$ as $t \to 0$. The form ω_t is of type $(1, 1)$ and satisfies $\partial_t \psi_t = \bar{\partial}_t \psi_t = 0$ so $d\psi_t = 0$. Let

$$\tilde{\omega}_t = \tfrac{1}{2}(\psi_t + \bar{\psi}_t).$$

Then $\tilde{\omega}_t$ is a closed $(1, 1)$-form which is positive definite for small t since $\tilde{\omega}_0 = \omega_0$ is positive definite. Thus $\tilde{\omega}_t$ is a Kähler form on M_t for small $|t|$.

Q.E.D.

Bibliography

Adler, A., The second fundamental forms of S^6 and $\mathbb{P}^n(\mathbb{C})$. *Amer. J. Math.*, **91**, 657–570 (1969).

Benzecri, J. P., Variétés localement affines. *Seminaire Ehresmann* (May 1959).

Borel, A., and J.-P. Serre, Groupes de Lie et puissances réduits de Steenrod. *Amer. J. Math.*, **75**, 409–448 (1953).

Bott, R., Homogeneous vector bundles. *Ann. Math.*, **66**, 203–248 (1957).

Calabi, E., and E. Vesentini, On compact, locally symmetric Kähler manifolds. *Ann. Math.*, **71**, 472–507 (1960).

Dieudonné, J., *Foundations of Modern Analysis*. New York: Academic Press, Inc., 1960.

Douglis, A., and L. Nirenberg, Interior estimates for elliptic systems of partial differential equations. *Comm. Pure Appl. Math.*, **8**, 503–538 (1955).

Ehresmann, C., Sur les espaces fibrés différentiables. *C. R. Acad. Sci. Paris*, **224**, 1611–1612 (1947).

Fischer, W., and H. Grauert, Lokal-triviale Familien kompacter komplexen Mannigfaltigkeiten. *Nachr. Akad. Wiss. Göttingen Math-Phys.* **K1 II**, 89–94 (1965).

Frölicher, A., and A. Nijenhuis, A theorem on stability of complex structures. *Proc. Nat. Acad. Sci. U.S.A.*, **43**, 239–241 (1957).

Gunning, R. C., and H. Rossi, *Analytic Functions of Several Complex Variables*. Englewood Cliffs, N. J.: Prentice Hall, Inc., 1965.

Helgason, S., *Differential Geometry and Symmetric Spaces*. New York: Academic Press, Inc., 1962.

Hironaka, H. An example of a non-Kählerian complex analytic deformation of Kählerian complex structures. *Ann. Math.*, **75**, 190–208 (1962).

Hirzebruch, F., *Neue Topologische Methoden in der algebraische Geometrie*. 2d edition. Berlin: Springer-Verlag 1962.

———,Über eine Klasse von einfachzusammenhängenden komplexen Mannigfaltigkeiten. *Math. Ann.*, **124**, 77–86 (1951).

Hirzebruch, F., and K. Kodaira, On the complex projective spaces. *J. Math. pure et appl.*, **36**, 201–216 (1957).

Hörmander, L., *Linear Partial Differential Operators*. Berlin: Springer-Verlag, 1963.

Kodaira, K., On the structure of compact complex analytic surfaces I. *Amer. J. Math.*, **86**, 751–798 (1964).

———,On cohomology groups of some compact analytic varieties with coefficients in some analytic faisceaux. *Proc. Nat. Acad. Sci. U.S.A.*, **39**, 865–868 (1953).

———,On Kähler varieties of restricted type. *Ann. Math.*, **60**, 28–48 (1954).

———,On compact complex analytic surfaces I. *Ann. Math.*, **71**, 111–152 (1960).

Kodaira, K., L. Nirenberg, and D. C. Spencer, On the existence of deformations of complex analytic structures. *Ann. Math.*, **68**, 450-459 (1958).

Kodaira, K., and D. C. Spencer, Groups of complex line bundles over compact Kähler varieties. *Proc. Nat. Acad. Sci. U.S.A.*, **39**, 868–872 (1953).

————,On deformations of complex analytic structures, I and II. *Ann. Math.*, **67**, 328–466 (1958a).

————,A theorem of completeness for complex analytic fibre spaces. *Acta. Math.*, **100**, 281–294 (1958b).

————,On deformations of complex analytic structures, III. Stability theorems for complex structures. *Ann. Math.*, **71**, 43–76 (1960).

Kuranishi, M., New proof for the existence of locally complete families of complex structures. *Proceedings of the Conference on Complex Analysis in Minneapolis 1964.* Berlin: Springer-Verlag, 1965, pp. 142–154.

Lang, S., *Introduction to Differentiable Manifolds.* New York: Interscience Publishers, 1962.

Milnor, J., *Morse Theory.* Princeton, N.J.: Princeton University Press, 1963.

Nakano, S., On complex analytic vector bundles. *J. Math. Soc. Japan*, **7**, 1–2 (1955).

Newlander, A., and L. Nirenberg, Complex analytic coordinates in almost complex manifolds. *Ann. Math.*, **65**, 391–404 (1957).

Palais, R. S., *et al.*, Seminar on the Atiyah-Singer Index Theorem. *Annals of Mathematics Studies No.* 57. Princeton, N.J.: Princeton University Press, 1965.

Serre, J.-P., Une théorème de dualité. *Comm. Math. Helv.*, **29**, 9–26 (1955).

Teichmüller, O., Extremale quasikonforme Abbildungen und quadratische Differentiale. *Abh. Preuss. Akad. der Wiss. Math.-naturw. Klasse*, **22**, 1–197 (1940).

Wu, W. T., Sur les classes caractéristiques des structures fibrées sphériques. *Actualités Sci. Indust., Hermann, Paris* (1952).

Index

Errata

Page v, line 9: "lines fundles" should read "line bundles".

Page 1, line -9: $c_{k...k_n}$ should read $c_{k_1...k_n}$.

Page 2, line 1 after the figure: There should be a comma between $|w_i|$ and i.

Page 2, line -5: $\leq$ should read $\subseteq$.

Page 3, lines 9, 11, and 13: $c_{k_1}\cdots k_n$ should read $c_{k_1\cdots k_n}$.

Page 3, line -3: $\partial \bar{z}_\nu$ should read ∂z_ν.

Page 4, line -13: $i \leq \nu \leq n$ should read $1 \leq \nu \leq n$.

Page 7, line -15: $z^u(p)$ should read $z^n(p)$.

Page 10, line 16: $\sum_{i=2}^{2n+1} x_i^2$ should read $\sum_{i-1}^{2n+1} x_i^2 = 1$.

Page 10, line 21: $v = 1, 2$ should read $\nu = 1, 2$.

Page 12, line 12: P^1 should read $\mathbb{P}^1$.

Page 12, line -5: $p \in G$ should read $g \in G$.

Page 13, line 2: F'_{m+s} should be F_{m+1}.

Page 13, line 2: g_m should read g, twice.

Page 18, line 10: Im $\omega \to 0$ should read Im $\omega > 0$.

Page 19, first line of the diagram: $\frac{1}{2}$ should be $\frac{\omega}{2}$.

Page 26, line 5: z_0^m should read z_2^m.

Page 32, line 13: $\tau_{\mu_1...\mu_q}$ should be replaced with $(k\tau)_{\mu_1...\mu_q}$.

Page 33, line -9: Omit the first $\mathscr{V}$ (in the subscript of Π). Replace the second script $\mathscr{V}$ with $\mathscr{U}$.

Page 33, line -5: Replace the first occurrence of $\Pi_{\mathscr{W}}^{\mathscr{U}}$ with $\Pi_{\mathscr{W}}^{\mathscr{V}}$.

Page 35, line 11: Replace all occurrences of $\mathscr{S}$ with $\mathscr{O}$

Page 38, line 5: $\eta_{\lambda\theta}$ should be $\eta_{\lambda\nu}$.

Page 39, line -10: M_0 should be M_b.

Page 41, lines 3 and 4: ≥ 1 should be > 1.

Page 42, line 10: z_1 should be z_2.

Page 43, line -6: ∂z_1 should be $\partial \zeta_1$.

Page 44, line 10: $-mg_{10}z_1 b_{10}$ should be $b_{10} - mg_{10}z_1$.

Page 55: Add the map Φ to the top arrow of the diagram.

Page 58, lines -13 and -6: $\in c^0$ should be $\in C^0$.

Page 61, line 3 after the diagram: $h\varphi$ should be $k\varphi$.

Page 64, line -13: $n \geq z$ should read $n \geq 2$.

Page 64, line -6: $f_{jk}(z) =$ should read $f_{ik}(z) =$.

Page 80, line -3: There should be a space after the comma between φ and q.

Page 81, lines 12 and 13: There should be a comma after 0; "The" should be "the".

Page 85, line -11: The last term in this equation should read $\sum_{\alpha,\beta} \frac{\partial^2 f}{\partial z^\alpha \partial \bar{z}^\beta} dz^\alpha \wedge dz^{\bar{\beta}}$.

Page 85, line -2: dz_α should read dz^α.

Page 86, line 10: $2\sum g_{j\alpha\beta}$ should read $2\sum g_{j\alpha\bar{\beta}}$.

Page 87, line -4: The second z in this equation is missing a subscript j.

Page 93, line -13: $\psi^{A_p\bar{B}_q}$ should be $\psi^{\bar{A}_pB_q}$; $\mu 1$ should be μ_1.

Page 93, line -9: $\bar{A}_q$ should be $\bar{A}_p$ both times.

Page 93, line -7: The first $\bar{B}$ in this equation should NOT have a bar.

Page 95, line -5: $\overline{\varphi \wedge *\psi}$ should read $\varphi \wedge *\overline{\psi}$.

Page 95, last line: The left-hand side of the last equation on this line should read $\delta\psi =$ instead of $\vartheta\psi$.

Page 100, line -10: The second equation on this line should read $\bar{\partial}f^\lambda_{jk\nu}(z) = 0$.

Page 108, line 12: omit the bar over ∂_λ.

Page 112, line -7: Proposition 5.4 should read "In the Kähler case ...".

Page 118, line 5: The last factor in the subscript of the R in the right-hand side of the equation should be ν.

Page 118, line 9: The subscript on the last R should be $\beta\bar{\nu}\lambda$.

Page 120, line -4: β_1 should be $\bar{\beta}_1$.

Page 120, last line: $(\bar{\tau})_i$ should be $(\bar{\tau})_k$.

Page 124, line 10: $\times$ should be $+$.

Page 126: line 5: β_q should be B_q; the term involving φ after the summation sign should be $\varphi_{jB_q} \cdot \varphi_j^{\overline{B_q}}$.

Page 126. lines 11, 12, and 13: All superscripts $\bar{B}_q$ should be B_q; in line 13 there should be a bar over the entire expression $\bar{\nabla}_\alpha\varphi_j^{B_q}$.

ISBN 978-0-8218-4055-9

About this book

This book, a revision and organization of lectures given by Kodaira at Stanford University in 1965–66, is an excellent, well-written introduction to the study of abstract complex (analytic) manifolds—a subject that began in the late 1940's and early 1950's. It is largely self-contained, except for some standard results about elliptic partial differential equations, for which complete references are given.

—D. C. Spencer, MathSciNet

The book under review is the faithful reprint of the original edition of one of the most influential textbooks in modern complex analysis and geometry. The classic "Complex Manifolds" by J. Morrow and K. Kodaira was first published in 1971 ..., essentially as a revised and elaborated version of a set of notes taken from lectures of Fields medallist Kunihiko Kodaira at Stanford University in 1965–1966, and has maintained its role as a standard introduction to the geometry of complex manifolds and their deformations ever since.

—Werner Kleinert, Zentralblatt MATH

Of course everyone knows Abel's exhortation that we should seek out "the masters, not their pupils," if we are to learn mathematics well and effectively. ... There is no question that this beautifully constructed book, full of elegant (and very economical) arguments underscores Abel's aforementioned dictum. Perhaps especially today, when so much is asked of the student of this material in the way of prerequisites, one can do no better than to turn to a master.

—MAA Reviews

The main purpose of this book is to give an introduction to the Kodaira-Spencer theory of deformations of complex structures. The original proof of the Kodaira embedding theorem is given showing that the restricted class of Kähler manifolds called Hodge manifolds is algebraic. Included are the semicontinuity theorems and the local completeness theorem of Kuranishi.

The book is based on notes taken by James Morrow from lectures given by Kunihiko Kodaira at Stanford University in 1965–1966. Complete references are given for the results that are used from elliptic partial differential equations.